The Southern
FOOD PLOT MANUAL

Ben H. Koerth
and
James C. Kroll
Institute for White-tailed Deer Management and Research,
College of Forestry,
Stephen F. Austin State University,
Nacogdoches, Texas 75962-6109

A Center for Applied Studies in Forestry Publication

First Edition
1 2 3 4 5 6 7 8 9

ISBN 0-938361-18-X

Publication also supported by:
Temple-Inland Inc.
Game & Fish Publications
North American WHITETAIL Magazine

Acknowledgments

While we put the words on paper, there is no way we can say we wrote this book by ourselves. There were a great number of people without whose involvement would have made this task impossible. First and foremost, we would like to thank our families for their unending support and patiently enduring our many hours away from home. Numerous edits and suggestions also made this book much better.

Second we would like to thank the current "deer crew" members Randy Browning, Mark Conway, Scott Dahler, Victor Donahey, Amy Kearns, and Tonya Sheffield for all their help in the office, lab and field.

Third, we owe special thanks to all of the individuals who graciously provided photographs: Drs. Tim Fulbright, Billy Higginbotham and Sim Reeves, as well as David Baggett, Ronnie Bane, Mark Conway, Bill Deauman and Richard Hancock.

Last, but certainly not the least, special thanks to all of the students who spent countless hours in the field helping with research projects and to all of the landowners and hunting clubs that allowed access to their properties.

Table of Contents

Introduction

Over the last decade, interest in whitetail hunting has increased to unprecedented levels. At the same time, the demand for quality hunting experiences has also escalated to impressive proportions. Hunting leases have become common in areas once thought to be strongholds of "open land" philosophy. Clearly, "these are the good ole days" for white-tailed deer hunting. But what about the future?

Deer populations have increased to record highs in most states, causing concern among white-tailed deer scientists and managers. Current estimates of whitetail populations in the United States exceed 26 million, with an annual harvest well over 5 million. Can we continue to see record harvests and monster bucks taken year after year in state after state? We think not. Many states now are showing signs of population decline.

In the near future, quality hunting experiences will be available only to those who

make an investment in the species. *Hunters hopefully will make the transition from consumers to managers.* The hunting experience will be far more than the mere killing of an animal. It will culminate in an active involvement of the hunter/manager in the lives and well-being of these magnificent creatures. Anti-hunters have accused us of using a euphemism for killing by saying we have *harvested* a deer. They are wrong! If we merely go to the woods, sit under a tree, and shoot a deer, we *have* indeed killed a deer. If, on the other hand, we have worked for many years improving the habitat, planting food plots, controlling the population and allowing young bucks to grow into their full potential, we truly can say we have *harvested* our buck. There is an important difference; don't forget it!

"There are no magic bullets! There are no magic beans!"

In many geographic areas there are significant site and soil limitations that prohibit a deer herd from realizing its potential. These site-limited areas *can* be improved to a productive level, but the average landowner or hunter often lacks the necessary information to make informed and effective decisions about augmentation of deer diets. *That is what this manual is all about—putting something back into the herd; through a sound, cost-effective management program.*

Herd Productivity <u>*IS*</u> Site Productivity

For too long now, some biologists have maintained there are areas that cannot produce quality deer. These areas, often referred to as *site limited,* are very common throughout the South and Southeast. "Landowners must learn to accept the limitations of the land," asserted one biologist. Yet, although it *is* true such sites cannot produce quality animals on their own, it *is not* true these same areas cannot produce better quality bucks *and* have higher productivity.

The problem lies in fertility of the site. For years now,

southeastern cattlemen have dealt effectively with this same problem; producing many times more beef per acre than their arid land counterparts to the west. For, the curse of the South (viz., abundant rainfall) also is our blessing.

Excessive rainfall, when coupled with vegetative cover like pine monocultures, can change basic soil fertility significantly. In many areas, heavy rainfall and pine plantations have produced soils with high acidity, resulting in low availability of important nutrients such as nitrogen, phosphorus and potassium. Consequently, plants growing on these soils--with production that can be measured in tons per acre--still provide suboptimum nutrition. *The problem is not quantity but quality!* Even preferred and highly palatable plants like blackberry, honeysuckle and greenbrier may fail to provide nutrition above maintenance level.

Deer are wild animals; we don't have to tell you that. Because they *are* wild animals, they are fully capable of surviving under a broad range of conditions. Even if native plants are suboptimum, deer manage to reproduce and live their lives quite well; as they have done for thousands of years. In order to accomplish this, however, they adapted in many ways to local site

limitations. In Florida, for example, deer seldom have more than spikes as yearlings or even two-year-olds. Fawn production is very low. While deer living on more productive sites exhibit a compensatory response to population control; you can shoot does until "the cows come home" in some Florida ranges, without any increase in fawn production.

Yet, in these same areas, cattle are produced in large numbers with high per acre yields. What is the difference? *The answer is: productivity management*. Long ago ranchers learned, in order to produce high yields of beef from their lands, they had to augment soil fertility and subsequently forage quality. Many of the improved forages used throughout the country were developed for grazing in Florida on poor sites.

When properly applied, the same forage management strategies can be used successfully to produce quality deer. However, the operative phrase here is "properly applied." For much too long now, hunters and landowners tried to supplement the diet of deer with limited success. Poorly educated as consumers, hunters easily fall prey to commercial claims. *There are no magic bullets! There are no magic beans!* There *are*, however, management strategies that, if properly implemented, can significantly improve *your* deer herd. We have written this manual to provide a common sense, step-by-step approach to food plot and supplemental forage management for whitetails. We emphasize management. There are no short cuts offered; just twenty-five years of experience.

Chapter One

The Basics of Supplemental Feeding

The first question is: Do you have a nutrition problem, requiring supplementation? Let's examine four important points.

1. What is the function of supplemental feeding?

It is easier to answer this question by discussing what supplemental feeding *is not*. Supplemental feeding is not replacement feeding. It is not intended to take the place of native forage; rather the purpose is to augment native forage quality and seasonal availability. *Supplemental feeding also is not a substitute for population control.* Quality whitetails only occur where herds are well beneath the carrying capacity of the range.

The truly monster bucks taken over the last decade all have come from areas with low deer populations. The Milo Hanson buck (net score 213 1/8 B&C points) taken near Biggar, Saskatchewan, in 1993 was from such an area. In Canada winter kill guarantees a herd that is considerably smaller than the fertile plains farm country will support. Because of low deer populations, every deer is assured of having more than enough to eat. It is only the periodic deep snow cover that kills these deer, not the cold.

Therefore, if you are not controlling your deer herd through adequate antlerless harvest, *and* you do not live near Biggar, Saskatchewan, there is no reason to read beyond this point. This book is not for you! On the other hand, if you live within the normal limits of the whitetail's range, and you *are* actively involved in a population management program, please continue.

Supplemental feeding, primarily through food plots, *can* significantly improve the nutritional plane of your herd. By so doing, you *can* improve fawn survival, body growth and antler development. In other words, just like the cattleman, you can greatly improve the quality and productivity of your herd.

2. You have a nutritional problem when...

When is supplementation necessary? Every state has biologists whose job it is to know the state's deer herd and habitat conditions. They can give you a lot of information about your specific area. Other management professionals such as Soil Conservation Service and state agricultural agents also are good sources of information.

In case professional help is unavailable, we suggest you investigate your deer herd and habitat. First, if you are doing a good job of management, you no doubt keep records on your deer herd and the animals harvested. Important records such as dressed weights, age-specific antler quality, incidental sightings and lactation rates (percentage of does in milk) all tell a great deal about what is happening nutritionally with your herd. [For a more in depth discussion on deer herd

Food plots, recommended in the South, provide supplemental forage at about one-tenth the cost of commercially prepared feeds.

management, refer to *A Practical Guide to Producing and Harvesting White-tailed Deer*, by Dr. James C. Kroll.]

If, for example, you have harvested large numbers of does *without* a corresponding increase in fawn production and survival, this may be an indication of site-limiting factors at work. Recently, several studies were presented at the Southeast Deer Study Group meetings documenting the lack of compensatory response to deer herd reductions in site-limited areas such as Florida, North Carolina and Georgia. A compensatory response means that you observe an increase in fawn survival when you reduce the number of deer on your land. In such areas, intensive habitat improvements and herd reductions may have little or no effect on herd productivity, body size or age specific antler quality. As suggested above, another clue would include a high spiking rate among yearling and older bucks, in spite of population control measures.

Deer utilization of supplemental forage often coincides with nutritional demands. Highest use occurs during winter and summer stress periods.

---

3. Food plots are the best way to supplementally feed in the South.

Basically, there are two ways to supplementally feed deer— feeding pelleted rations and food plots. Pelleted rations are the most costly, ranging from $0.30 to $1.60 per deer per day. We use pelleted rations only in areas where food plots are not practical and/or as a backup to food plots during drought periods. Food plot forages cost about $0.01 per pound, making the cost of supplemental feeding via food plots about one-tenth the cost of commercial feeds.

4. Food plots are important even when there is *not* a nutritional problem.

Often, even in areas with adequate forage quality, supplemental food plots may be important. Basically deer managment can be classified into two philosophies. First, there is *extensive management* under which you attempt to manage the herd within the

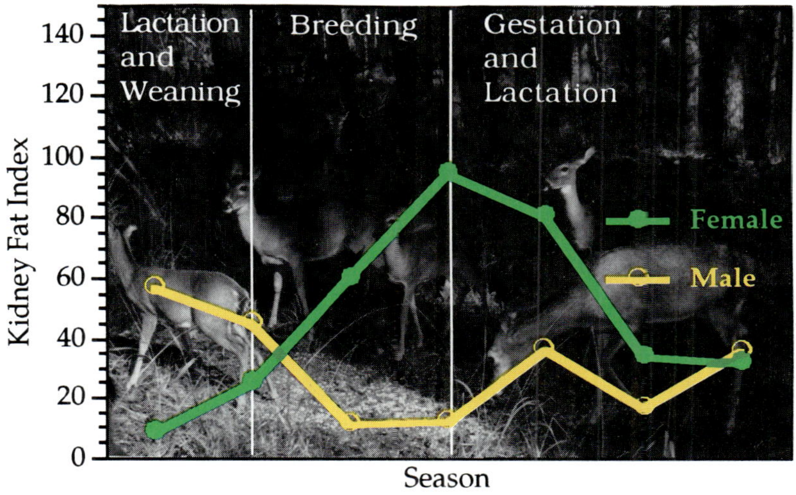

Generalized kidney fat curves show a seasonal response to reproduction and obligatory fat deposition in the fall. (From Johns et al. 1984, Journal of Wildlife Management).

---

limitations of the habitat. In an *intensive management program*, on the other hand, both the herd and habitat are manipulated to produce at a higher level of productivity. Supplemental feeding can enhance fawn production, body size and antler quality of bucks above and beyond typical extensive management programs. For example, with the deer herd at Fort Perry Plantation near Columbus, Georgia, we have supported extremely large-bodied animals (250+ pounds), high fawn crops (130%+) and extremely high quality antlers using an intensive year-round food plot program.

Supplemental Food Plot Strategies

White-tailed deer annually experience one or more stress periods. In the South, there are primarily two stress periods. Research has shown the first occurs in mid- to late-winter, when high energy foods such as acorns and other mast are low or virtually depleted. Acorns have a high content of

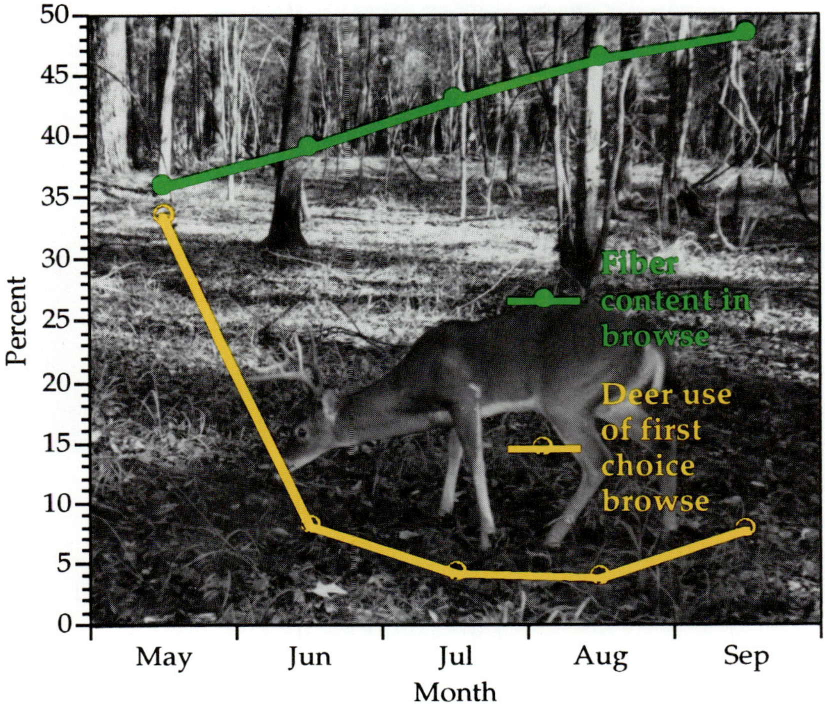

Use of native browse plants decreases as the fiber content increases with maturity because this makes plants less digestible. (From Bane 1992)

---

fat and carbohydrates, but have a relatively low protein content. Extended cold periods put a heavy demand on fat reserves; and, how well deer can handle these stress times is often determined by the amount of body fat stored in early fall. Hunters and landowners often argue their herds are not over-populated because the deer harvested during the hunting season have stored fat. The amount of fat in an animal harvested in fall belies its true condition. Biologists prefer to make summer collections of deer to determine health and body condition. At this time, the amount of kidney and abdominal fat gives a truer testimony to relative stocking level and habitat quality. Fat deposition, called lipogenesis, in deer is obligatory; meaning they store fat in the fall even under poor forage conditions.

By late winter, fat stores often are fully exhausted, leaving the animal to survive or die according to weather conditions. The amount of reserve fat also affects reproductive success the following spring and summer. In some areas, deer herds are "mast driven"; meaning years with good acorn crops are followed by high fawn crops, while poor years produce few fawns. Examples of these "boom and bust" areas of the country are the Hill Country of Texas and the Piedmont of Georgia.

White-tailed deer have been conditioned by eons of natural selection to reduce protein requirements during the cool season. Elaborate recycling systems allow whitetails to reduce their need for protein intake. Therefore, deer must deal primarily with energy limitations in winter. Summer is another story.

The second stress period occurs in late summer or early fall. This is a time when most plants have ceased growth in

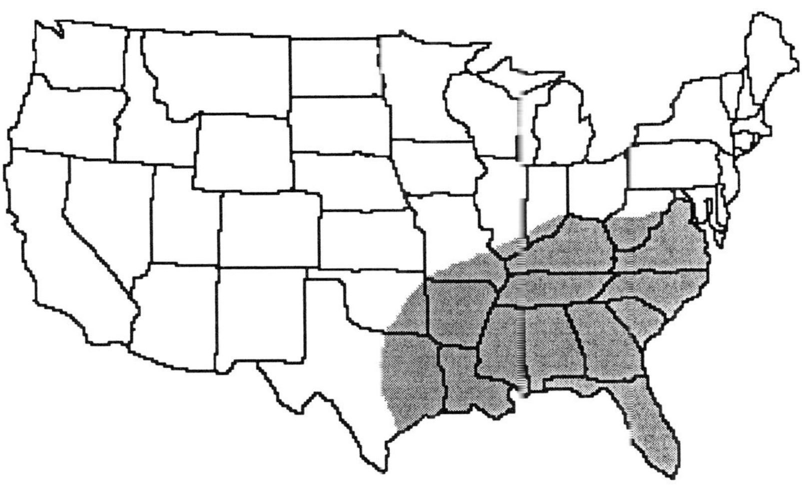

While many plant species mentioned in this text can be grown successfully in other areas, the primary area of concern is indicated by the shaded portion of this map.

response to low rainfall and/ or typical plant growth habit. Plants that are growing generally are higher in protein and minerals, and are more digestible than mature plants. Asparagus provides a good example from a human perspective. Asparagus is tasty only during active shoot growth. Once growth is complete, the plant deposits woody fibers. The result is an unappealing and indigestible meal! So it is with native deer forages.

Although there are species, such as honeysuckle, which continue to grow throughout the summer, most deer forage has a distinct growing season, after which it becomes less palatable and nutritious. Protein content of native forages drops well below maintenance levels (10-12% protein) by mid-July in most southern ranges.

Hence, in the South we must deal with two stress periods; each placing entirely different demands on the deer. The winter stress period is characterized by the lack of adequate energy; while the summer stress, although also somewhat energy limiting, is characterized by protein and mineral deficiencies. In order to effectively mitigate these deficiencies, you must adopt a *management strategy* that deals both with cool- and warm-season stress periods. After more than twenty years of forage research, we came to the conclusion no single plant species can do this! *To be effective, you will have to plant mixtures of species and make more than one planting each year.*

The basic theme of this manual will be *supplementation strategies*. In the past, too little attention has been paid to this subject in the published literature. Implementation of these strategies will mean a commitment in time and money. An effective food plot program is hard work! But, we feel the rewards are worth the cost. Again, we offer a caveat. *If these hard facts cause you concern, then supplemental plantings probably are not for you. If you are committed to giving something back to the deer, read further.*

Chapter Two

Getting Started

Even beginners with little or no experience can successfully establish supplemental food plots for wildlife. A little common sense and a few basic guidelines, however, can help make your venture more successful. Whenever it's possible, food plots should be planted in existing openings to reduce establishment costs. It's also a good idea to get aerial photographs or a map of your area to aid in site selection. Examples of good locations to establish food plots are natural meadows, firelanes, logging trails, pipeline rights-of-way, logging decks, fallow fields or grassy lanes, and sides of interior roads. When you're hunting, scouting, practicing your outdoor skills—anytime you are in the woods—look for candidate areas. Mark these areas on your aerial photos or map of the area. Also include notes about size, general soil condi-

tions, probability of flooding, amount of sunlight, accessibility and amount of deer sign. If you plan to hunt the plot, scout for likely stand locations. After you have catalogued all available areas, your list of potential food plots can be ranked and the area or areas you plan to establish as food plots can be matched to the funds, equipment and manpower available.

To provide adequate nutrition to cover the 2 major stress periods for deer in the South, it is necessary to plant both warm- and cool-season supplemental food plots. Frankly, if you exclude warm-season varieties from a food plot program, you will have significant limitations in your management program. In many cases, warm- and cool-season crops can be developed in the same plots, especially with no-till planting methods discussed later. One mixture is simply planted over the existing vegetation without interrupting the availability of your forage. If you have the land area available, you may desire to establish separate plots to plant warm- and cool-season varieties. Another example would be to divide a single plot, with a cool-season mixture in one portion and a warm-season mixture in the other.

Where to Plant

Other things to consider when planting any food plot are site characteristics such as size, soil type, moisture and sunlight. Let's discuss a few things about each of these aspects.

1. Location

Location of your food plot will determine limits on size, what soil type you will be planting in, as well as inherent soil moisture conditions and amount of sunlight that can reach the plants. However, several other factors also may come into play. Placement of food plots is critical to success. Locate your food plots where deer will feel secure in using them. It can be very frustrating if you try to attract animals long distances from escape cover. Locate your plots near travel corridors, natural feeding areas, or other habitats that are frequented by deer. If your plots are too far from adequate concealment, there is the danger that deer will use them only under the cover of darkness, if at all.

2. Shape

Shape of your food plot is not particularly important and can vary tremendously. Many miles of pipeline, utility line and road rights-of-way are already in place over much of the country. These types of areas readily lend themselves to establishing supplemental food plots. In most cases they have easy access and are already cleared of trees and other woody vegetation. Rights-of-way are typically long and narrow in shape. Long, narrow plots usually are preferred because they are easily established and increase the amount of edge available compared to square plots.

"Locate your food plots where deer will feel secure in using them."

Irregularly shaped plots that follow the contour of natural openings further increase the amount of edge and are more aesthetically appealing. Scattered trees or other cover maintained in the plot adds to the natural appeal. However, in cases where livestock are present, food plots should be fenced to maximize the production of forage for deer. If fencing is needed to exclude livestock, then a square is the most efficient design to minimize fencing cost. We have found two or three strands of electric wire will turn away cattle without limiting deer access. Other fencing, such as barbed wire, net wire or livestock panels work well, but are more expensive than electric fence.

3. Distribution and Size

Size of food plots is important. Whitetails, being an "edge species," are more apt to feed along edges of cover than in large open areas, so several smaller plots are preferable to one large plot. A good rule of thumb is to have plots a maximum of 100 yards wide. Deer are reluctant to venture much beyond this distance in daylight hours. Normally, the recommended size for food plots is at least one half acre and up to 3 acres. Plots of at least one half acre are large enough deer are not likely to over-graze your plot in a short time and move to other areas. Plots larger than 3 acres, unless designed as long and narrow, can result in portions of the plot being too far from cover and not being used by deer. Remember, ten 1-acre plots distributed across the habitat are preferable to 1 plot 10 acres in size; and uses the same amount of seed and fertilizer.

A good recommendation is to plant from 1 to 3 percent of the habitat in food plots, with both warm- and cool-season plants. In other words, 1 to 3 acres should be established for each 100 acres of habitat, with at least one plot established in each 160 acres of available habitat.

Landowners and hunting clubs can team up with rights-of-way owners to the benefit of both. Pipeline and utility line companies must keep the rights-of-way cleared of trees and brush. This can lead to considerable annual expense. In some cases, rights-of-way owners have provided seed, fertilizer and lime; the landowner or hunting club provided the labor and equipment to plant the right-of-way. Establishing and maintaining the food plot will keep the right-of-way clear of

In 1990, a first-of-its-kind program was initiated by David Baggett, Wildlife Biologist for Champion International, hunting clubs that leased the land and Gulf States Utilities. Basically, a formal contract was entered whereby Gulf States furnished the seed, fertilizer and lime, and the hunting club planted and maintained the clover under the administration of Champion. For the cost of mowing and lightly disking the clover stand each year, the hunting club received high quality supplemental food plots. The utility right-of-way owner benefited from substantially lower or eliminated expense of mowing or spraying to keep the right-of-way cleared of unwanted woody vegetation. The program was highly successful and has expanded considerably. Check with your local utility and rights-of-way owners for similar cooperative programs.

unwanted brush. If this is an option available to you, try it. Everyone benefits including wildlife.

4. Soil

It is important to consider soil type when establishing food plots. Deep sandy soils tend to be excessively drained and will not have adequate moisture available for warm-season plants except during rainy periods. Sandy-loam type soils typically are well-drained but retain adequate moisture for plant growth except during drought periods. Clay soils typically occur in bottomland situations and tend to retain more moisture than upland locations during dry weather.

One of the top three mistakes commonly made in establishing supplemental food plots is ignoring soil fertility. Take the time to collect a soil sample, have it analyzed and amend your plot with the recommended nutrients to *assure* successful supplemental food plots. Further information on this aspect will be discussed later

5. Light

No plant can exist without light. Sunlight is necessary for a plant to photosynthesize and produce food for itself. However, the amount of sunlight each plant requires differs depending on its particular growth habit. Some plants require full sunlight while others do best in partial or complete shade. Plant species recommended for food plot establishment normally develop and thrive on 3 to 4 hours of sunlight each day. When establishing your food plot, pay close attention that your plot is not excessively shaded by nearby trees. Trees also tend to reduce available moisture under the dripline of their crowns.

Soil Tests

Soil has a direct effect on the kind and amount of vegetation available that can be grown as food, as well as cover for wildlife. The proper amount and balance of nutrients like nitrogen, phosphorous and potassium are necessary for optimum plant growth. Soil pH, a measure of acidity, largely determines the ability of plants to extract these essential soil nutrients through their root system. As different plants thrive on various levels of nutrients and pH, a soil test is paramount in determining what element(s) may be deficient.

1. Soil Sampling

Bags or cartons for soil samples along with detailed instructions usually can be obtained from your County Agricultural Extension agent, the local Soil Conservation Service office or from the nearest soil testing laboratory. Sometimes fertilizer dealers provide soil sample bags.

To take a soil sample:

- *Divide the fields where you intend to establish food plots. Never mix different types of soils in the same sample.*
- *With a spade, trowel or other small digging tool, make a small, vertical hole in the ground in 10 to 15 different places within each field. Take a thin slice of soil that is free of vegetation from the backside of each hole and place this in a clean bucket or other container for mixing. It's very important the mixing container is clean and free of contaminants*

and chemicals.

- *Break up the slices you placed in the bucket and mix thoroughly.*
- *Place about a pint of this mixed soil in the sample bag or carton.*
- *Label each sample carefully so you know the plot it came from and can match the soil laboratory recommendations to the correct plot. Be sure to indicate on the soil sample bag or attached information sheet the type of crop(s) you desire to establish (e.g., warm-season legumes).*
- *Lastly, be sure to complete all of the supplementary information about cropping history and past fertilizer applications. This in-*

Use a small shovel (a) to take a thin slice of soil from several locations within the field. Place the slices into a clean bucket (b) and mix thoroughly. Put about a pint of mixed soil into the sample bag (c). Lastly, be sure to carefully label sample bag (d) and completely fill out all supplementary information.

formation is critical and may provide as much information as the actual soil test in determining the fertilizer requirements.

Soil samples should be taken several months before you intend to plant. This gives the soil testing laboratory time to complete the analysis and determine the best option to properly amend your soil for the particular crop you want to establish. Also, if lime is recommended to adjust soil pH, it may take a month or more to become completely available to the plants after application. After lime and fertilizer needs have been determined, soils should be amended on an annual basis.

2. Analysis and Interpretation

Most state agricultural universities have laboratories that perform soil tests for a nominal fee. If you are not familiar with a soil testing facility in your area, the office where you picked up the sample bags or cartons can provide you with a location of the one nearest you. A list of soil testing laboratories is included in Appendix II at the end of this book.

The soil test report will provide nutrient values and pH. Nutrient values may be shown in general terms (e.g., low, medium, high) or in more specific terms as parts per million or in pounds per acre. The general levels are more important because these indicate the probability of a significant plant response with the addition of that particular nutrient. A low nutrient level in the sample would indicate a high probability of plants responding to that nutrient being added to the soil by fertilization. Conversely, a high nutrient level would indicate little response can be expected by adding that nutrient.

Without getting technical, a breakdown of the basic soil nutrients may be helpful in understanding. Seventeen nutrients are recognized as *essential elements* for plants because normal growth depends upon their presence. Three of these essential elements, carbon, hydrogen and oxygen, are obtained primarily from the air and cannot be amended in a practical sense. Of the remaining fourteen elements, six are used by plants in relatively large amounts and thus are called *macronutrients*. These include nitrogen (N), phosphorus (P), potassium (K), calcium (Ca), magnesium (Mg) and sulfur (S). Typically, concentration levels of most macronutrients are included in all soil test reports.

Nitrogen, phosphorus and potassium are commonly supplied in commercial fertilizers and, consequently, are known as the *primary* or *fertilizer* elements. Sulfur, magnesium and calcium also are amended with fertilizers and are called *secondary* elements. Magnesium and calcium are most often added to soils in limestone. Sulfur is normally applied alone or as an ingredient in fertilizers like manure, superphosphate or sulfate of ammonia.

The other eight elements — iron, manganese, copper, zinc, boron, molybdenum, chlorine and cobalt —are used in very small amounts by plants; thereby deriving the name *micronutrients*. This does not imply they are less essential, but merely used in smaller quantities. Concentration and ratio of micronutrients are as critical to proper plant growth as are macronutrients. An imbalance of micronutrients is likely to occur on heavily leached acid soils, highly alkaline soils, muck soils, soils that have been intensively cropped and fertilized with macronutrients only, or where land leveling or similar operations have removed the topsoil and you are planting crops in the subsoil.

Concentration differences at which micronutrients are deficient and where they can be toxic to plants are extremely small. In addition, adding macronutrients in commercial fertilizers can be expensive. Therefore, making fertilizer recommendations is no job for an amateur. It would be wise to place your confidence in personnel experienced and trained in local soil conditions who use your soil test data and supplementary information of past land uses and fertilizer applications, along with a complete understanding of principles underlying field procedures.

Based on analysis of nutrient levels along with soil pH, soil test reports will give the minimum recommended fertilizer requirements for establishment of the crop you indicated. If you have questions about the report, your County Agricultural Extension agent, Soil Conservation Service personnel, or a reputable fertilizer dealer are

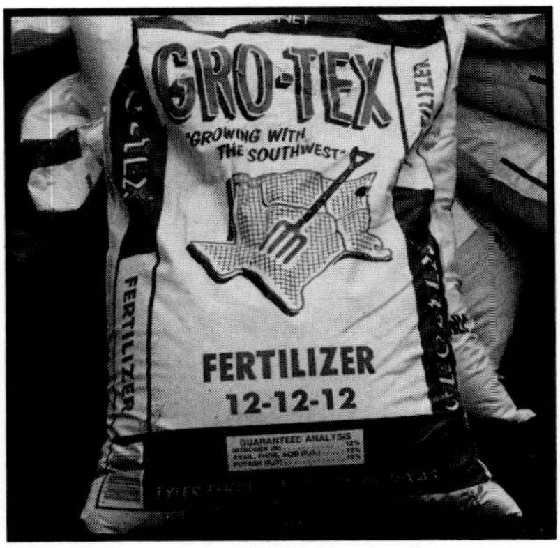

very knowledgeable and can help you interpret the results.

3. Fertilizer Components

By law, all commercially produced fertilizer material must carry a guarantee as to its content of the fertilizer elements; that is, nitrogen, phosphorus, and potassium. The exact form may vary by state. Nitrogen usually is expressed simply as nitrogen (N), phosphorus as available phosphorus (P) or phosphoric acid (P_2O_5), and potassium as water-soluble potassium (K) or potash (K_2O). In the simplest form of analysis guarantee, the numbers on fertilizer bags indicate the relative levels of N, P, and K, or N, P_2O_5 and K_2O in the fertilizer mixture. The higher the number, the more of that particular nutrient is included. For example, an 8-16-16 fertilizer contains 8 percent total nitrogen, 16 percent available P_2O_5 (7 percent available P), and 16 percent water-soluble K_2O (13 percent water-soluble K). Nutrient values can be zero. For example, on a phosphorus-poor soil that contained an adequate amount of nitrogen and potassium, you might see a recommendation for a fertilizer like 0-46-0.

Don't expect to run down to your local dealer at the last minute and buy a bag of 8-24-24! Plan ahead and have the dealer custom mix your formulation for you. If this is not possible, try to make adjustments with fertilizer combinations available. For instance, fertilizers can be grouped by the nutrient ratio. A 4-8-8, a 5-10-10, and a 10-20-20 all have a 1-2-2 ratio. Accordingly, one ton of a 5-10-10 fertilizer furnishes the equivalent N, P and K as one half ton of 10-20-20.

Now you have your soil report and it looks something like the sample report at the end of this chapter. In this example, the recommended minimum fertilizer treatment for sample no. 1 is to apply 0 lbs/

acre of N, 40 lbs/acre of P_2O_5 and 60 lbs/acre of K_2O. The question is what fertilizer do you use and how much will you need to achieve the recommended rate? First determine how many lbs/acre of total fertilizer mixture you want to apply; somewhere between 200 and 400 is the norm. If you use less than 200 lbs/acre it becomes difficult to distribute the fertilizer evenly, particularly over small plots. For an example, let's use 400 lbs/acre as our fertilizer rate.

In this case, nitrogen is easy. Don't apply any nitrogen to warm-season legumes as nitrogen tends to enhance growth of native grasses that flourish in direct competition with your planted species. The fertilizer you want will have 0 as the first of the three numbers.

For phosphorus, divide the recommended rate from the soil report by the total amount of fertilizer you want to apply per acre and multiply by 100. That is, the soil report recommends 40 lbs/acre of P_2O_5 and we want to apply 400 pounds of total fertilizer mixture per acre, so:

$$(40 \div 400) \times 100 = 10.$$

Now we know the first two numbers in the fertlizer we want is 0-10. The same calculations are performed for K_2O. That is:

$$(60 \div 400) \times 100 = 15.$$

Therefore, in this case we want to apply 400 pounds of 0-10-15 for each acre of food plot associated with soil sample 1.

That was easy. However, different levels of nutrients are recommended for samples 2 and 3 in the soil report. Do you have to change fertilizers? In instances where the fertilizer recommendations are radically different between samples, the answer would be yes. In that case you would recalculate the fertilizer requirements as shown above. However, remember the ratios discussed earlier. In the example soil report, notice the fertilizer recommendations for all three samples have a 0-2-3 ratio; the same as our fertilizer. The difference is samples 2 and 3 require one and a half times as much P_2O_5 and K_2O as did sample 1. We use the same fertilizer mixture but we need to apply half again as much total amount per acre. The result is:

Measuring Land by Pacing

An alternative method of land measurement is the ***chain****. This is a very old measurement, dating back to colonial times, and is equal to 66 feet. There are 80 chains in a mile (5,280 ft ÷ 66= 80); 10 square chains in an acre; and 43,560 ft^2 in an acre.*

Chains are measured easily by pacing. You can establish ***your*** *pace by laying out a distance of 66 ft and walking at a normal gait along this distance. Remember, however, a* ***pace*** *is two steps. So, start out each time with your left or right leg, counting each time you return to that leg. For most men, the number of paces in a chain ranges from 10 to 13. You should not take great big strides; rather, walk at your normal gait.*

Repeat this at least three times and take the average number of paces per chain. You now have "calibrated" yourself for making land measurements.

In measuring distances, count the number of paces to the nearest half pace. Then divide this number by your pace to obtain the distance in chains. For example, suppose you average 12 paces per chain. You measure a distance to be 60 paces. Divide 60 by 12 to obtain 5 chains or 330 ft (5 x 66).

---

400 x 1.5 = 600 lbs/acre of 0-10-15.

4. Soil pH

Soil pH can range from 1 to 14 and is a measure of the acidity of the soil. A level of 7 is considered neutral, less than 7 is acidic; more than 7 is alkaline. Each change of 1 unit is considered a tenfold increase or decrease in acidity. For example, pH of 4 is ten times more acidic than pH of 5 and 100 times more acidic than pH of 6.

Extreme soil pH values normally range from about 3.5 to 10.5. Soils in humid regions, characteristic of the Southeast, normally range from about 5.0 to around 7.0. In other words, moderately acidic to neutral. Soils in more arid regions char-

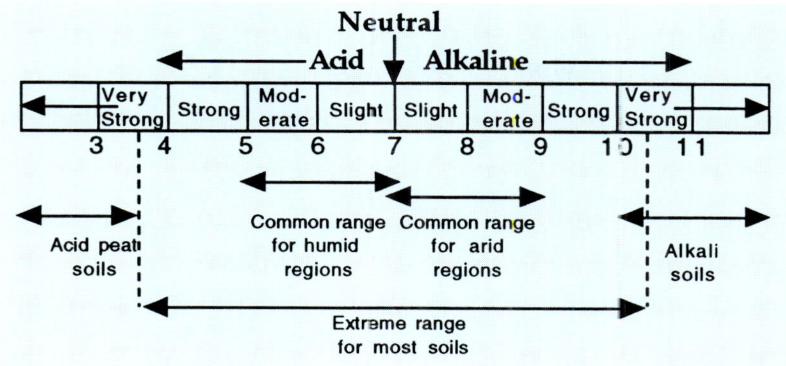

Soil acidity, or pH, determines the availability of many nutrients for plant growth.

---

acteristically range from moderately alkaline to neutral.

Soil pH determines the availability of nutrients for use by plants. As pH goes above about 7.5, some essential nutrients like iron, zinc and manganese will become less available. At pH values less than about 5.0, aluminum, manganese and iron can become so soluble that they may be toxic to some plants. Phosphorus, while never readily soluble, becomes the most available around a pH of 6.5. Phosphorus availability is very low when pH falls below about 4.0, or is above about 7.5 to 8.0. When necessary, soil pH can be adjusted with the addition of lime (to increase pH from acidic) or sulfur (decrease pH from alkaline), in addition to change induced by fertilizer ingredients.

In all cases, a routine soil test provides an economical method to ensure the proper nutrients are available in the correct proportions.

Determining Size of Plot

We often ask a landowner who is seeking advice on supplemental plantings about the size of his food plots; and, we get some pretty strange answers. Unfortunately, most people do not know the size of their food plots. "Oh, it's about an acre." On measurement, however, it turns out to be one-third of an acre. Obviously, applying fertilizer, lime and seed based on an inflated estimate could lead to some serious problems, as well as wasted money. You should obtain a reasonable estimate of the area of each food plot. This often is not easy, particularly if the plots are irregular in shape. But, there are some good methods which produce adequate results.

The easiest food plot to measure is one that either is square or rectangular. Using the formula, **A= L x W**, you can obtain the area in square-feet of your plot. In this case, L is the length of the plot and W is the width in feet. Since there are 43,560 ft^2 in an acre, you divide the area by this figure to obtain the acreage. Here's a example to explain what we mean:

Length of plot = 200 ft

Width of plot = 100 ft

Area = 200 x 100 = 20,000 ft^2

Acreage = 20,000 ÷ 43,560 = 0.46 acre

For irregularly shaped plots, it is a bit more difficult; but, not impossible. The best way (see illustration) is to establish lines vertically and horizontally across the plot. You either can measure or pace these lines to get an average length and average width. Pacing really is the best way to do this. Here's an example using measured distances.

Suppose you have the irregular plot shown in the illustration. There are 5 vertical lines and 8 horizontal lines to measure. [The number is arbitrary; selected in this case to best fit the situation.] The total length for all 5 vertical lines is 750 feet; the total for all horizontal lines is 1,580 feet. Hence, the average length is 750 ÷ 5, or 150 feet; while the average width is 1,580 ÷ 8, or 197.5 feet. Now, multi-

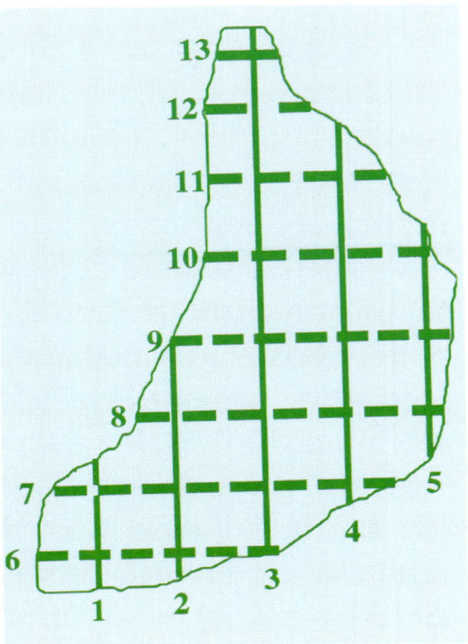

ply the two (150 x 197.5) to obtain an estimated area of 29,625 ft^2 in the plot. The final step is to divide the area by 43,560 to obtain 0.68 acre.

If you pace distances, you can use area estimates calculated as square chains. A square chain is equal to 4,356 ft^2, so 10 square chains equals one acre (43,560 ft^2). Simply determine the average length in chains, multiply this figure by the average width in chains, to yield area in square chains. Then divide by 10 to calculate acreage.

An alternative method would be to obtain an aerial photograph of the area, showing the food plots, and tracing around each plot with a planimeter, or by over-laying a dot grid. Soil Conservation Service and extension agents all know how to use these methods and may be willing to help.

Calibrating Your Spreader

Proper calibration of spreaders is of the utmost importance to achieve effective and economical application of seed and fertilizer. On small plots you might get by with a light application rate and simply going over the plot several times until all of the material has been spread. However, on large plots or where several plots are to be planted in succession, this is very inefficient. Without calibrating, you also run the risk of over-seeding or fertilizing at first and running out of material before you finish.

The simplest method of calibrating is trial and error. First determine how wide a swath your spreader covers. Then measure the distance needed for a area of about one-tenth of an acre using the formula for area given earlier in this chapter. Weigh enough seed or fertilizer to plant one-tenth acre and pour it into the hopper of the spreader. Apply the material while walking or driving, depending on the equipment, at the speed you intend to plant. When you reach the end of the measured distance, determine amount of material used. From that you can increase or decrease the amount of material dispensed as necessary. Repeat the entire operation until your desired flow rate is achieved.

Power takeoff- or hand-operated machinery can be calibrated using a stationary method. Measure a calibration area as above. Determine the amount of time necessary to cover the area. This should be an average of several trips across the area to be treated. Fill the hopper with enough seed or fertilizer to cover the calibration area and attach a collecting device onto the output of the spreader. Operate the spreader as if you were planting for exactly the time it takes to cover the calibration area. Increase or decrease the amount of material dispensed until the desired flow rate is achieved. In other words, when all of the material is used in exactly the time necessary to cover the calibration area.

Record-Keeping for Food Plots

As with all aspects of deer management, it is important to maintain records on all activities. Remember, a short pencil is better than a long memory. You should keep a log (see example) showing *all* activities performed on *each* food plot. Each plot should be numbered and illustrated on a map or aerial photograph of your land. Current aerial photographs can be obtained from one of several

Aerial photographs can be ordered from the following locations.

Region 2 (Western Mapping Center)
National Cartographic Information Center
U.S. Geological Survey
345 Middlefield Road
Menlo Park, CA 94025

Region 3 (Rocky Mountain Mapping Center)
National Cartographic Information Center
U.S. Geological Survey
Box 25046, Stop 504
Denver Federal Center
Denver, CO 80225

Region 4 (Mid-Continent Mapping Center)
National Cartographic Information Center
1400 Independence Road
Rolla, MO 65401

Region 5 (Eastern Mapping Center)
U.S. Geological Survey
536 National Center
Reston, VA 22092

U.S. Geological Survey offices around the country (see addresses at the end of this chapter). If aerial photographs are not available for your area, or are too old to include your food plots, you have two alternatives.

First, you can rent a small airplane, usually for less than $100, and take your own photographs. You will need a 35mm camera, preferably with a 135mm lens. In most cases, you should photograph your area from an altitude that allows you to include your entire property in one or two photographs. Second, you can draw an outline of your food plots on old aerial photographs using their location in relation to known features of the landscape.

Each year, you should schedule all activities, recording the planned date of completion on your food plot log form. Give yourself a "window of opportunity," because weather or unforeseen circumstances may delay action.

Detailed records should be kept on all soil test results, lime and fertilizer application rates, seed varieties and seeding rates, planting methods, and any observations pertinent to the plot. These records permit better planning and can aid in troubleshooting should a problem arise.

SOIL TESTING LABORATORY
STEPHEN F. AUSTIN STATE UNIVERSITY

Soil Test Report

Name Michael L. Morton-Lavaca Creek Hunting Club Address 2101 S. Fredonia, Nacogdoches, TX County Nacogdoches Date 3-30-94

SOIL TEST RESULTS AND CLASSIFICATION*

Lab No.	Sender's Sample No.	Soil pH	Buffer pH	Nitrate Nitrogen NO_3-N ppm	Phosphorous P ppm*	Phosphorous P level	Potassium K ppm*	Potassium K level	Calcium Ca ppm*	Calcium Ca level	Magnesium Mg ppm*	Magnesium Mg level	Salinity hazard	Sulfate Sulfur SO_4-S ppm	Iron Fe ppm	Manganese Mn ppm	Zinc Zn ppm	Copper Cu ppm
64740	1 CR at Tram	6.5	N.A.	1	10	L	173	M	1605	H	486	H	VL	H				
64741	2 W. Tram	6.0	7.65	1	1	VL	56	VL	1011	H	160	H	VL	M				
64742	3 Camp Area	5.6	7.52	1	2	VL	69	VL	403	M	147	M	VL	H				

*The values for ppm (parts per million) are based on testing procedures used at SFA Soil TestingLaboratory and may differ from results obtained by other procedures. The level of very low, low, etc. is more important than ppm, since these levels indicate the probability of response to added plant nutrients. A very low level indicates a high probability of response to added plant nutrients. A very high level indicates there is little chance of response to added plant nutrients. VL=Very Low; L=Low; M=Medium; H=High; VH=Very High; > Means-More Than; < Means-Less Than

SUGGESTED FERTILIZATION AND LIMING RATES

Sample No.	Crop	Limestone Tons of ECCE/Acre	N (pounds of plant nutrients per acre)	P_2O_5	K_2O	REMARKS
1	Various Warm Season Legumes	0	0	40	60	Minimum requirements to establish warm season legumes.
2	Various Warm Season Legumes	.6	0	60	90	Minimum requirements to establish warm season legumes.
3	Various Warm Season Legumes	1.0	0	60	90	Minimum requirements to establish warm season legumes.

If the limestone applied contains ≥ 3% magnesium it will help build Mg levels, but this Mg content is not required.

If you have questions about this soil test report, see your County Agricultural Extension Agent, Soil Conservation Service Personnel, or your Fertilizer Dealer.

Recommended by Terry L. Bell

The upper portion of this soil test indicates pH ranged from slightly to moderately acidic in these three soil samples. A low to very low level of phosphorus was determined in all three samples. Potassium levels were classified as moderate in sample 1 and very low in samples 2 and 3. This indicates a good probability of significant plant response to the addition of these two fertilizer elements. Fertilizer recommendations given in the lower section reflect the different levels of phosphorus and potassium found in the soil and the amount of lime necessary to sufficiently amend pH. Notice no nitrogen fertilizer is recommended for planting warm-season legumes. Refer to the text for determining kind and amount of fertilizer to meet the recommendations.

HISTORY OF AREA SAMPLED

1. Plot name/number______________________

2. Position (check one)
   - ☐ 1st bottom
   - ☐ bench
   - ☐ upland
   - ☐ greenhouse
   - ☐ muck or peat

3. Internal drainage (check one)
   - ☐ sluggish
   - ☐ imperfect
   - ☐ normal
   - ☐ excessive

4. Slope (check one)
   - ☐ depression
   - ☐ flat
   - ☐ gently rolling
   - ☐ rolling
   - ☐ steep

5. Intended use (check one)
   - ☐ permanent pasture
   - ☐ long-time meadow
   - ☐ cropland rotation
   - ☐ lawn
   - ☐ other ____________________________

6. Has the area been limed in the past 4 years?______Yes ______No

If so, when?
- ☐ this year
- ☐ last year
- ☐ 2 years ago
- ☐ 3 years ago

What rate?
- ☐ less than 1 ton/acre
- ☐ 1 ton
- ☐ 2 tons
- ☐ 2 or more tons

7. If fertilizers or manure have been added in the past 4 years.

Year	Crop Grown	Fertilizer Added		Tons Manure per acre
		Rate (lbs/ac)	Analysis	

Example of supplementary information useful to a soil testing laboratory to accurately interpret your soil analysis and make recommendations for fertilizer. History of land use and past fertilizer treatments affect soil fertility significantly. The type of crop you intend to plant also will influence soil amendments.

SUPPLEMENTAL FOOD PLOT RECORDS

Management Unit______________________________ Year___________

Plot Name/Number_________________________ Size (acres)_________

Planting Information

Crop to be planted:____
Seeding rate:_______________ Depth:________ Date:______________
Planting method:______________ Moisture conditions:_____________

Crop to be planted:____
Seeding rate:_______________ Depth:________ Date:______________
Planting method:______________ Moisture conditions:_____________

Crop to be planted:____
Seeding rate:_______________ Depth:________ Date:______________
Planting method:______________ Moisture conditions:_____________

Fertilizer Information

	N	P_2O_5	K_2O	Date
Fertilizer application (lbs/ac):	_____	_____	_____	__________
Top-dressing (lbs/ac):	_____	_____	_____	__________

Lime (tons/acre)___________________ Date:______________________

Soil Analysis

Date last sampled:_______________________

Soil Lab Analysis

pH	P	K	Mg	S

Fertilizer Recommendation

Lime	N	P_2O_5	K_2O	Mg	S

Other Activities

Herbicide:______________ Rate:_____________ Method:______________
Insecticide:_____________ Rate:_____________ Method:______________

Chapter Three

Combinations Are The Key

Food plots will be of little benefit to your deer if your plantings do not provide *what* the deer need or are not available *when* the deer need them. Plants are *not* created equal, at least in a deer's eye. Unfortunately, we have found no single forage species that can provide a completely balanced nutritional diet for deer. Also, food plots are not used solely by deer. Any number of non-game, as well as game species, including dove, quail, turkey, rabbits and many others, benefit from plantings made by hunters. Insectivorous birds and bats actively seem to seek out food plots for foraging because of the large number of insects attracted to the lush green plants, particularly warm-season plantings. Turkeys use food plots as brood habitat. For these reasons and a number of others mentioned later in this

chapter, mixed plantings of two, three or more species of plants are recommended.

Why Combination Plantings?

Different plants typically do better on some sites than on others. Combination plantings increase the probability at least some of the plant species will do well in each plot you plant and reduce the risk of losing a crop to insects, disease or adverse weather. In addition, different plant species grow and mature at different rates. As a species matures, another is available to emerge as the primary producer. Ultimately, your goal is to provide a continuous supply of diverse and highly digestible food.

If you are establishing a food plot for the first time, it is a good idea to make small plantings of several different plant species and varieties in order to determine which plants produce best in your area. *Don't rely on magazine advertisements and marketing claims.* Make your own test plantings on different soils in bottomlands and uplands. By the end of the first year, you will have narrowed your choice of plants considerably and can

Combination planting of alyceclover and American jointvetch.

make an informed decision about future plantings.

Although new research may produce even better plant varieties, we caution you, "*do not* jump on every new plant species or variety you see mentioned in a magazine or advertisement." The species and varieties discussed and listed in this handbook are "tried and true." Stick with them and they undoubtedly will reward your efforts. We often are amused by someone who, in spite of great success, switch to a new, untried planting apparently in a never ending search for the "holy grail" of food plots.

Matching Plant Combinations and Sites

Remember, different plant species are adapted to various soil conditions. Some plants grow best in well-drained soils in drier upland sites. Others prefer moist soil conditions, and usually do better in bottomland sites. Few, if any, of the plants most often recommended for food plots for deer will survive in areas that are flooded.

Warm-season plants often do best in bottomland sites, particularly if these areas are not flood prone. Bottomland sites usually retain more moisture during drier summer months than uplands. Good candidate areas for planting include level areas around ponds, creeks, streams and temporary drainages. An important criterion in selecting warm-season plants is they must be productive enough to compete with native grasses and weeds.

Cool-season plants, on the other hand, do not have to contend with hot, dry weather conditions found in summer. Competition from native grasses and weeds also is not a major factor. These plant varieties can be planted either in bottomland or upland sites.

Before selecting a species or variety for your food plots, be sure your choice can meet these criteria:

1. *production of enough forage to justify establishment costs,*
2. *readily eaten by deer,*
3. *nutritional value in terms of percent crude protein and digestibility that are suitable for deer,*
4. *available during the stress periods when native forage is*

lacking in quality, and

5. *able to withstand heavy grazing pressure.*

Additionally,

6. *cool-season species should be cold tolerant, and*
7. *warm-season species should be drought tolerant.*

Warm-Season

Supplemental food plots typically have been aimed primarily at aiding harvest of deer, with little thought given to improving nutrition during summer. Consequently, cool-season plant species have been the prominent type of food plots. Some hunters seem to think if they can't hunt over it, it's wasted. Although it's often overlooked, in the South, summer can be more of a stress period than winter. Another important point is much of the South has a long growing season with a late frost. Warm-season food plots are still producing and attracting deer well into the hunting season. This is important, especially to the bowhunter, as it seldom frosts during bow season.

During the hotter, drier parts of summer, most native plants mature or become dormant. Remember, plants that are actively growing are the most nutritious. However, this is the time of year when deer need a high protein diet the most. Does are giving milk, fawns are growing and being weaned; and, maybe most importantly, bucks begin

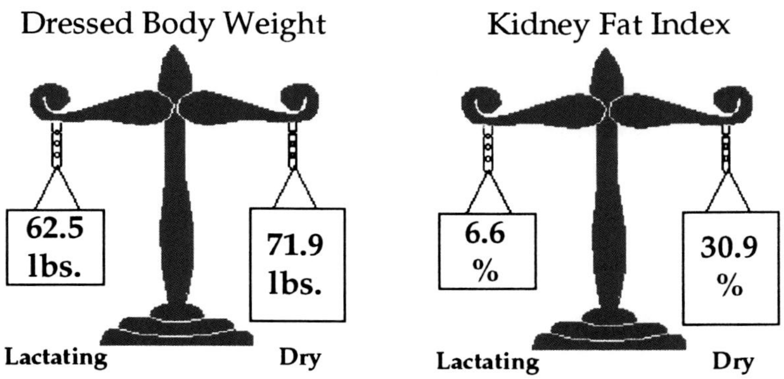

The cost of raising fawns is demonstrated by lower dressed body weight and reduced kidney fat in lactating does. (Kroll and Koerth, unpublished data)

growing the antlers we covet so much later in the season.

Some of the most commonly planted warm-season varieties are field peas including red ripper, black-eye, purple hull, catjang, and iron and clay cowpeas. These forages have the potential to produce highly nutritious forage well-liked by deer. However, *when planted alone, unless on food plots several acres in size, field peas will not produce an adequate food plot for deer.* In dry years, grazing pressure from the deer never allow the peas to produce significant growth. In more favorable years, the peas might produce a lot of forage, but are wiped out by deer in a short time. Deer are like kids and candy when it comes to field peas. They tend to feed on peas even when adequate native forage is available. For a food plot to be effective, it's critical that the forage be available when the deer need it the most.

Summer clover and clover-like legumes in combination with field peas appear to make the best warm-season food plots. A comprehensive study of warm-season food plots in the South by Drs. Billy Higginbotham and James C. Kroll found the forages that performed the best included alyceclover, iron and clay cowpea, catjang cowpea, American jointvetch and centennial soybean. Plantings that included alyceclover and cowpea performed the best in upland sites. Because American jointvetch thrives in moist soil conditions, mixed plantings of alyceclover, cowpeas and American jointvetch produced best in bottomland sites. Alyceclover and American jointvetch also withstood heavy grazing pressure much better than the pea varieties. The ability to recover from grazing and continue to grow is an important point in keeping a continuous supply of forage available.

Cool-Season

Cool-season plantings that combine cereal grains with legumes have produced the best results. Arrowleaf clover (Yuchi, Meechee, Amclo varieties), ryegrass, cold-tolerant varieties of oats and rye do well in well-drained sandy-loam to clay soils in uplands. Bottomland sites,

however, are usually too wet to produce good stands of these species. In bottomlands, a mixture of white clovers (especially ladino) and wheat has produced with good success.

Most clovers typically are slow growers initially. Cereal grains in the mixtures provide early growth and production to attract deer. Clovers then emerge to provide quality nutrition later as the cereal grains begin to mature. Again, this provides the continuing cycle of productive plants to supplement the diet of deer.

However, if different plant species have different growth rates, how can we use this to benefit not only the deer but the deer hunter as well? We recommend planting a higher proportion of cereal grains to clovers on food plots that are hunted because of their attractiveness to deer early in the season. Assuming normal growing conditions, cereal grains germinate and grow much earlier than do clovers. This provides an early season "magnet" that draws deer to your hunting spots. Clovers typically germinate later and grow slowly for the first few weeks. However, clovers are there to ultimately provide the high quality nutrition to the deer herd during the stress period later in the winter when the cereal grains are maturing.

What do you think would happen if your roadside plantings or other nonhunted food plots were as attractive during hunting season? The deer would be scattered all over the place, which would make hunting and management success much more difficult. Chapter Nine in this manual discusses this in further detail.

Typical supplemental forage combinations.

Crop Species	Site	Seeding* Rate (lbs/ac)	Planting Depth
Warm-season 2 species combination			
Cowpeas	Bottomland	40	1/2-1"
Alyceclover	Bottomland	10	1/4-1/2"
Warm-season 3 species combination			
Cowpeas	Bottomland	40	1/2-1"
Alyceclover	Bottomland	10	1/4-1/2"
American Jointvetch	Bottomland	5	1/4-1/2"
Cool-season 4 species combination			
Rye	Upland or Bottomland	75	1/2-1"
Oats	Upland or Bottomland	25	1/2-1"
Arrowleaf clover	Upland or Bottomland	10	1/4-1/2"
Ryegrass	Upland or Bottomland	10	1/4-1/2"
Cool-season 5 species combination			
Rye	Upland or Bottomland	60	1/2-1"
Oats	Upland or Bottomland	10	1/2-1"
Wheat	Upland or Bottomland	20	1/2-1"
Arrowleaf clover	Upland or Bottomland	10	1/4-1/2"
Ryegrass	Upland or Bottomland	10	1/4-1/2"

* Seeding rate is for broadcast. Reduce rate by 20 percent if seed drill is used.

Chapter Four

Planting Your Food Plots

One of the first decisions you will have to make in planting food plots, after you decide what to plant and where to plant it, is whether to use till or no-till planting techniques. We have found there is a decided reluctance among hunting clubs and landowners to use no-till or low-till methods. There seems to be little faith these methods will work. Also, there is a mistaken tendency to think of wildlife food plots as agricultural crops that have to be maintained weed-free. Remember, native weeds are some of the best wildlife food plants and can be as nutritious as the crop you're planting. Because, as we have noted, no single technique or plant variety works in all situations, we will discuss both methods.

Basics of Till versus No-till

To begin, let's define the

The three basic methods of planting include *till* (upper left) where the soil is plowed or broken, *no-till* (upper right) where the seeds are planted without cultivation, and *low-till* (left) where seeds are drilled directly into the soil but the soil is not plowed and existing vegetation remains relatively undisturbed.

difference between till and no-till planting. *Till methods are any form of cultivation where the soil is plowed or broken in preparation for sowing seeds.* The purpose is to remove all vegetation and leave a bare soil surface upon which to plant. Farmers and gardeners have successfully used this method for countless years.

Conversely, *no-till means planting without cultivation of the soil*—basically, the same way seeds are planted in nature, although the aid of modern machinery is often used. In its simplest form, seeds are broadcast onto the ground where they germinate, grow and compete for soil nutrients and water along with existing native vegetation. In other cases, seeds can be drilled into the soil, but the soil surface is not turned or broken and existing vegetation remains relatively undisturbed. This method of planting is called *low-till.*

Advantages of tilled plots: In theory, the maximum production from the planted crop would be realized from tilling the soil because all competing vegetation is removed. This leaves the soil and fertilizer nutrients and all available moisture for the intended crop. In addition, turning undecomposed green plant material into the soil is a form of fertilization often referred to as *green manuring*. This brings about a number of favorable benefits for future crops including incorporation of organic matter into the soil, addition of nitrogen if plants are legumes, and recycling of nutrients that are stored in the plants. Furthermore, removing vegetative cover and tilling of the soil may expose some insects to cold or heat, birds, or other enemies that can aid in their control.

Disadvantages of tilled plots: Inherent in this method is the need for mechanized equipment and considerable energy input. Probably the most common equipment used to prepare food plots would be a tractor and tandem disk. Disturbing the soil hastens the drying process, possibly leaving seeds to the mercy of drought if a rainfall event does not follow soon after your planting. Exposed soil also is prone to erosion problems, particularly in food plots that are disked flat with no furrows or terraces to stop runoff. How many times have you seen a food plot with the topsoil, as well as seeds, washed to the bottom of the slope?

Tilled plots also require separate areas be established for warm- and cool-season plantings. This necessitates twice the tillable land area unless one crop is plowed under in preparation for planting the next. In this case, the deer are left without a supplemental forage supply until the new crop attains enough growth for grazing. As we have noted previously, the most effective supplemental planting program assumes an even supply of quality forage throughout the annual cycle.

Advantages of no-till: With no-till you can establish food plots in remote areas not accessible by machinery. This method allows planting both warm- and cool-season crops on the same area without interrupting the food supply. As the cool-season plants mature and go dormant, warm-season plants are there to emerge and become the dominant producers for the coming growing season. A similar cycle is repeated for cool-season plantings in autumn.

After no-till planting, it often is advantageous to mow, bushhog or sling down tall grasses and weeds to allow germinating seeds to compete more successfully with the existing vegetation. This also leaves a mulch on top of the seed that preserves soil moisture and protects the seedlings from being grazed at a very early stage. Stems and roots of existing vegetation are left to control erosion.

Disadvantages of no-till: Because the vegetation that is established on the food plot remains undisturbed, the species you plant needs to grow quickly and be able to compete aggressively. This becomes particularly important for warm-season crops because native vegetation will be enhanced as well by fertilizer applications. This necessitates being an early bird and getting your food plots planted *before* the native plants attain enough growth to effectively choke out new seedlings. Also, if the crop you planted last season is still producing and has formed a dense stand that will limit germination of new seedlings, it may have to be mowed. While mowed plants will still produce forage for grazing, seed production may be limited, thereby reducing the potential of the stand to reseed itself. *Timing* is the key to success in no-till planting.

Preparing the Site

1. Soil Amendments

If you haven't already done so, the first step in site preparation is to have the soil tested. This applies to both till and no-till planting methods. As is continually repeated throughout this manual, a laboratory analysis is the *only* accurate way to determine which soil nutrients may be limiting plant growth. For optimum performance, follow the recommendations for fertilizer and lime fully.

Fertilizers and lime can be

The appropriate inoculant must be used with each plant species selected. First mix the inoculant with the seed (a), apply a sticking agent (about 12 ounces of cola per 50 pounds of seed (b) or a commercial sticking agent as prescribed in the label directions) and mix thoroughly (c). Plant the seed as soon as the sticking agent dries sufficiently for the seed to disperse properly (d).

---

applied in a number of ways, but the most common method is to broadcast with a truck or tractor-mounted spreader. If you intend to till the plot, the fertilizer and lime should be disked into the soil as the seedbed is prepared. This places the fertilizer in the best position to serve the seed that is planted. If the plot is just being established, it may be necessary to bushhog brush and heavy vegetation and subsequently disk the area several times. The goal is to disturb the soil thoroughly and remove all of the surface vegetation.

If you use no-till planting methods, fertilizer and lime are applied broadcast but are not worked into the soil. Again, this places the fertilizer in close proximity of the seed.

2. Site Preparation

Tilled plots will require the most site preparation. After the fertilizer and lime have been incorporated into the soil, plants with large seeds like cowpeas or cereal grains can be broadcast and disked into the soil either along with or after the fertilizer has been applied. Do not disk small-seeded plants, like clover, into the soil as this will cover them too deep to get proper germination. Instead, the seedbed should be firmed and leveled. A roller, cultipacker or light drag works well for this operation. After the seed is broadcast onto the plot, the roller or drag is run over the plot to place the seed in firm contact with the soil.

For no-till planting, seeds should be broadcast onto the plot following fertilization and liming. In most cases this will be all that is necessary.

A home-made drag used to cover small-seeded plants like clovers.

No-till and low-till methods have the advantage of not interrupting the food supply as new seed is planted directly over existing vegetation or mulch left after mowing.

However, if existing grasses and weeds are over about 8 inches tall and dense enough that the ground is not readily visible as you walk through the plot, you need to mow, bushhog, weed-eat, or sling down the existing vegetation. The purpose is to stunt the native vegetation to give your food plot a head start. This step can be omitted if vegetation is very low, sparse, or is going dormant. We like to establish a bahiagrass sod through which we no-till. Bahiagrass forms a virtual single-species sod that is open enough for easy no-till or low-till planting. Close grazing or mowing often is desirable for maintaining a good stand of bahiagrass.

Planting Instructions by Forage Class for Tilled Plots

1. Cereal Grains

Most cereal grains have relatively large seeds and can be disked into the soil as deep as 1 to 1-1/4 inches without having a negative effect on germination. For seeding rates, refer to planting tables in Chapters Three and Ten. Rye is a good choice to plant in a cool-season forage combination because of its cold hardiness. Other grains that complement rye include oats and wheat, although rye should make up two-thirds of the combination. Winter wheat is resistant to cold. Oats, on the other hand, are susceptible to freezing temperatures and are easily killed if exposed. If oats are to be used in the planting combination in a region that receives heavy freezes, be sure to use a cold-tolerant variety.

Steps in planting a food plot include taking a soil sample (a), amending the soil according to the soil test (b), inoculating the seed properly with the correct inoculant (c) and carefully planting the seed (d) according to recommendations. Maintaining a healthy stand necessitates follow-up maintenance and continual monitoring against possible troubles.

2. Legumes

Inoculation of all legumes with nitrogen-fixing bacteria is mandatory before planting. Because inoculant varies with the species being planted, be sure to use the appropriate inoculant. Mix the inoculant thoroughly with the seed. It is best to use a sticking agent to assure the inoculant adheres well to the seed. Add about 12 ounces of soft drink or commercial sticking agent per 50 pounds of seed and mix thoroughly again.

Legumes with large seeds like cowpeas or cereal grains can be disked into the soil to a depth of about 1 inch. Small-seeded legumes, like clover and vetch, should not be disked, but rolled or lightly dragged. Combination plantings may require different planting techniques. Let's take a typical warm-season combination planting of cowpeas and alyceclover for example. The cowpeas are first broadcast and disked. Next, the seedbed should be firmed and leveled as described above. The clover is broadcast over the same area and lightly dragged or pressed into the soil with a roller or cultipacker. A small piece of chainlink fence works nicely if you don't have this equipment. Small-seeded varieties should never be planted more than about 1/4 to 1/2 inch deep for best results.

No-till strategies

As we have noted before, the basic strategy behind no-till farming for deer is to minimize soil disturbance. Soil disturbance not only increases moisture loss in warmer months, but also stimulates weeds that compete with the planted crop. Any method that reduces soil disturbance also will minimize these two negative effects.

There are two basic strategies to no-tilling your food plot. First, you can broadcast directly over the native vegetation. This is followed by mowing to reduce competition and lay down a mulch to conserve moisture. Second, you may want to use a new technique discovered by seed producer Richard Hancock (Wildlife Seed Blend, Inc., Dade City, FL). In this method, a permanent sod is established using a bunching grass such as bahia. Bunch grasses, unlike varieties such as bermuda, have a large amount of bare ground between each plant. This has proven to be an excellent situation for top-seeding with food plot varieties, *without* disturbing the soil or plants. The sod then can be left intact or mowed as discussed above. We have tried this technique along roadsides and rights-of-way, with excellent success. This also is a good technique when soil erosion is a major concern.

Intermediate Fertilizer Treatments

In all cases, proper fertilization and liming at the time of establishment is paramount. However, to realize the true production potential of cool-season plots, follow-up applications of high nitrate fertilizers are usually necessary. Cool-season food plots should be top-dressed with 200 pounds per acre of a high nitrogen fertilizer like 34-0-0 at about 60 days after planting.

A special note about fertilizers and legumes is in order. Summer plantings of legumes, especially no-till or low-till, should not be fertilized or top-dressed with high nitrogen fertilizer. Remember, because legumes fix their own nitrogen, and do not really need this nutrient, applications of high nitrogen fertilizers only encourage competition from other plants, especially grasses.

Food Plot Maintenance

Once you have established a food plot, common sense dictates you want to implement a maintenance program that minimizes both effort and cost. No-till certainly allows this, but there are some cost-effective strategies for till as well. Unless you over-graze your plot, you should have adequate seed production from legumes the first or second year. You may be asking, "How does one over-graze with wild animals?" Simple, by allowing the deer herd to remain well above carrying capacity. As we noted at the beginning, if you are not making a serious commitment to population control, all the food plots in the world will not benefit your deer. If you plant less than the recommended 2 percent minimum area, or inadequately harvest your herd, your food plots will not be able to reseed.

However, we will assume you *are* managing your herd properly and you *have* planted the minimum acreage both in warm- and cool-season plots. In that case, you should have an adequate seedbank to produce a solid stand of clover next season. Clovers readily reseed, however, legumes such as cowpeas and soybeans do not reseed normally and have to be planted each year.

Let's take a look at two different ways to maintain permanent cool- and warm-season plots. Then we will discuss strategies for using the same plot for over-lapping plantings of warm- and cool-season varieties.

Maintenance of cool-season plantings: Cool season plantings are the easiest to maintain. In most cases, varieties such as arrowleaf, crimson and red clovers produce a heavy seed crop during spring to early summer. During a wet, cool summer, arrowleaf often persists well into July in many southern habitats. Where it's applicable, red clover may persist all summer. Other clovers such as crimson produce seed much earlier, usually during the period March-May. Whatever the variety, large amounts of seed lie beneath a mat of dead vegetative matter. Once the plants produce seed and die back, native plants continue growing well into the fall, providing additional nutrition resulting from the residual fertilizer and nitrogen fixed by the clover.

Unless cool-season plots have to double for warm-season plantings as well, it is a good idea to not disturb these plots until fall. The only summer activity should be acquisition of soil samples for establishing amendments for the fall planting. Lime should be added during the summer, if required, to allow time for incorporation into the soil. Fertilizers can be added prior to planting.

There are two approaches to fall food plot maintenance. First, you can apply needed fertilizers and topseed with cereal grains (excluding oats) and/or ryegrass. Next, mow the plot as close to the ground surface as possible. This produces a layer of organic matter on top of the seed. Finally, *lightly* disk the plot so that most of the seed is covered by no more than 1/2-inch of soil. Light disking may cover legume seeds too deeply, but does leave a large amount of this seed lightly covered. Hence, light disking can produce satisfactory results. You would want to use this approach if you are planting oats in the plot, since oats are not suited for top-seeding.

Using no-till and low-till approaches, however, require slightly different methods. In no-till planting, apply fertilizer as above, broadcast seed, then simply mow the existing vegetation to provide a mulch over the seed. In low-till, mow first, then drill through the mulch.

Maintenance of warm-season plantings: Warm-season plantings can be maintained similarly. They too can be tilled, no-tilled or low-tilled, depending on the situation. In tilling, you should follow the basic procedures outlined for cool-season plots. *Remember not to use nitrogen fertilizers, however.* Large-seeded legumes such as peas and soybeans should be disked in lightly. We conducted research on re-establishing small-seeded, warm-season legumes such as alyceclover, jointvetch and hairy indigo. Both mowing or *light* disking produced acceptable results, followed by no treatment and heavy disking, respectively.

In no-till and low-till maintenance, you should broadcast or drill large-seeded legumes as given above for cereal grains. Timing of mowing and planting also should be as given above for cool-season plots.

Overlapping warm- and cool-season plantings: Often, there are situations in which you cannot afford the luxury of having separate cool- and warm-season plots. But, you also do not want to have a prolonged period in which your deer do not have adequate nutrition. There is a way to overlap warm- and cool-season plots, with a minimum loss of forage production.

Our research has shown a cool-season planting that consists of an early maturing clover such as crimson, along with the standard cereal grain combination, will allow no-till or low-till planting to warm-season plants in late spring. If you plant a clover variety such as arrowleaf that maintains significant growth well into early summer, however, it is not possible to use this technique. We like to drill right through the grain and dead clover, then mow the plot. Mowing lays down a mulch that helps maintain soil moisture for seed germination. Also, mowing retards maturity of the cereal grains and residual clover.

Again, as with all warm-season no-till and low-till procedures, timing is of paramount importance. You should plant as soon as soil temperatures reach 68-70° F. Earlier plantings are possible, but may lead to a reduced crop or even failure if seeds are exposed on the soil surface for too long before conditions are right for germination.

Summary

There are no shortcuts to proper food plot establishment and maintenance. A good food plot program begins with proper selection of site and plant varieties, then considers soil amendments and proper planting procedures. We have found the most common reason for crop failure is improper planting. You should follow these simple rules to success:

1. *soil test;*
2. *add appropriate amount of lime and fertilizer;*
3. *inoculate all legumes;*
4. *plant at the correct time and by the correct method;*
5. *provide for a sound plot maintenance program; and*
6. *monitor plot conditions and health.*

Chapter Five

Measuring Production, Utilization and Quality

If supplemental food plots are to successfully meet your management goals, then it matters a great deal how much forage is produced, how well the forage is accepted by the grazing animals and if it effectively meets their dietary needs.

Exclusion Cages

In some cases, if deer are using a food plot heavily, it is difficult to determine just how much forage you have produced. In other instances, if production is good and your food plot is of adequate size, it may be difficult to determine how much forage the deer are utilizing. In all cases, you should construct a small exclusion cage in each of your food plots, or at the minimum, the food plots that are easily

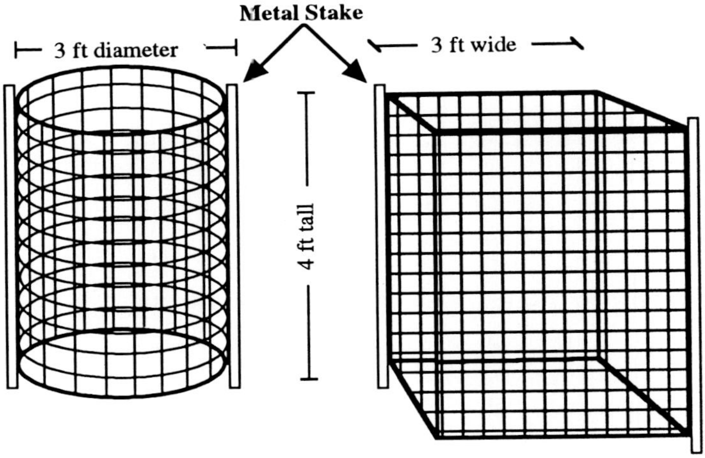

Round or square exclusion cages can be easily constructed from 2 x 4 inch mesh netwire and metal t-posts. Cages remain the best method to determine production and use of food plots by deer.

observed on a regular basis. Cages are best made from any type of net wire with a mesh size small enough deer and other animals cannot get to the forage inside the cage. Exclusion cages do not need to be big and can be of any shape. A cage about 3-4 feet square or a round cage about 3 feet in diameter will suffice. All cages need to be about 4 feet high so deer cannot reach in over the top. Be sure to stake the cages well to prevent deer or other animals, especially feral hogs, from pushing them around. Two-by four-inch welded wire securely tied to metal t-posts make excellent exclusion cages.

The amount of forage produced inside the cage, where it is not eaten, is a conservative measure of total production of your food plot. The difference in production between the inside and outside of the cage indicates the extent of utilization of the food plot.

Forage Sampling

High forage *production* is one essential element of a good supplemental food plot. The other is high forage *quality*. White-tailed deer depend on forage to supply all of their dietary needs including protein, carbohydrates and minerals. How well your supplemental forages meet the

White-tailed deer are concentrate feeders meaning they have the ability to select plants and plant parts that provide the best nutrition available.

nutritional requirements at a particular time of year is a measure of quality. The phrase "at a particular time of year" is important because nutritional requirements change with the specific activities taking place at that point in time (see Chapter One).

To determine if the forage you have available is meeting the dietary demands of deer, have the quality of the forages tested. Many farmers and ranchers routinely have similar tests conducted on the quality of their hay. This can be done at any forage testing laboratory, often located in conjunction with a soil testing facility (see listing in Appendix II).

1. Obtaining a Forage Sample

To obtain a sample, hand pick a representative sample of the forage available in the food plot. Remember, deer are very picky eaters and tend to select the tender parts of the plants. They also select from a number of different species if they are available. You want the hand-picked sample to represent as accurately as possible what the deer are actually consuming. The sample does not have to be very big; a large handful is usually adequate. Place the sample in an airtight and watertight plastic bag and place it in a cool location to retard bacterial growth that can alter laboratory analysis. Samples can be frozen for a short time until they can be delivered to the forage testing laboratory.

2. Analysis and Interpretation

Common forage analyses include crude protein level and a measure of digestibility, usually ADF (Acid Detergent Fiber) or NDF (Neutral Detergent Fiber). In general, the higher the crude protein content, the greater the nutritional value. In most cases the report will show 2 values for percent protein in the sample that may be quite different and cause some confusion. Although the exact wording will vary, basically the 2 values will represent percent crude protein on an *oven dry* basis and on an *as fed* or *as sampled* basis. Because all forages are made up of different percentages of water, the only way to make an accurate comparison of nutrient levels is for the laboratory to remove the water and then make determinations only from the dry matter. Your forage should be producing a minimum of 14-16 percent crude protein on a *dry basis* during the active growing season.

ADF and NDF are measures of the relatively indigestible, fibrous portion of the forage. Specifically, ADF values consist primarily of the amount of lignin and cellulose. NDF values are composed primarily of lignin, cellulose and hemicellulose; all of which are poorly digested by deer. Therefore, NDF values will be slightly higher than ADF. In both cases, the lower the ADF and NDF values the higher the digestibility and qualityof the forage. ADF values of less than 40 percent and NDF values less than 50 percent are the most suitable for deer.

Other tests you may want to have conducted include digestible energy and calcium and phosphorus content. Digestible energy is a measure of carbohydrate and fat content and is most important during fall and winter for 2 reasons. First, in the fall deer spend a great deal of time seeking high energy foods to build the fat stores necessary to survive the winter. Hunters often take advantage of this by hunting near mast producing areas. Second, bucks come out of the rut with badly depleted fat reserves. The availability of high energy foods may determine whether or not the buck you passed up last season survives.

Minerals, like calcium and phosphorus, are necessary for proper body functions and bone growth, including antlers. Calcium seldom is lacking in most areas, but phosphorus often is well below miminum requirements. In areas with high rainfall and acid soils, phosphorus is typically low. Optimum antler and body growth is obtained with relatively small quantities of calcium and phosphorus. Minimum requirement for calcium is about 0.45% and as low as 0.28% for phosphorus, however, the ratio of calcium to phosphorus in the diet is critical. A good planting should produce a 2:1 calcium to phosphorus ratio or better. Food plots remain one of the best ways to "input" these nutrients into your deer herd.

Mineral Supplementation

Hunters and managers are often interested in providing additional mineral as part of a supplemental feed program. There certainly is no harm in this practice as long as it is done correctly and provides extra assurance that your deer are receiving a complete mineral supplementation.

Hunters and managers have long recognized the attractiveness of natural salt licks. Deer readily use these areas and probably receive other mineral supplements from these deposits. If natural licks are not available in your area, you have the opportunity to provide mineral licks with commercially formulated mineral supplements.

As with planting food plots, you first have to understand that providing supplemental mineral is not a cure-all for deer management problems. It is simply one step in a complete management program. Second, while salt is a great attractant to deer, salt alone is not a balanced mineral supplement. As with humans, large quantities of salt are not healthy. A high salt content also means you are getting less of the other minerals that are necessary for a well-balanced supplement. Mineral should be limited to no more than 30-35 percent salt. Other criteria include a 2:1 or better calcium to phosphorus ratio and micronutrients including iron, potassium, magnesium, zinc, sulfur, manganese, copper, cobalt, iodine, selenium and vitamins A, D, and E.

There are several commercial manufacturers of mineral that are acceptable. To determine if the brands available in your area meet the requirements for deer, look at the contents label before purchasing the mineral to determine what is included.

Further, use of mineral blocks is not recommended. Deer seldom use hard blocks until they melt with rainfall and the mineral can be licked from the ground surrounding the block. Providing a loose or granular mineral is preferred.

Where the mineral is supplied is another important criteria. Some people pour a granular mineral on a stump, on top of the ground or dig a shallow hole to hold the mineral. While these practices may be necessary in some cases, you should realize any time there is concentrated feeding on the soil surface there is the danger of spreading pathogens and parasites. Mineral supplied in a covered trough is the preferred method. Troughs designed to hold supplemental feed (discussed in the next chapter) can be used to hold mineral as well.

In addition, if you are going to provide supplemental mineral, have it available all year long. There are times when minerals probably are more important than others, particularly when young deer are growing and bucks are beginning antler growth. If mineral is supplied *ad libitum*, deer can decide what, how much and when they want it.

STEPHEN F. AUSTIN STATE UNIVERSITY
SOIL TESTING LABORATORY
FORAGE TEST REPORT

Date: May 15, 1993

Name: Dr. James Kroll
Forestry Department
P. O. Box 6109
Nacogdoches, Tx 75961

County: Bosque Lab #: 11014

Type of Hay: Ryegrass and Clover Sample #: Sodd Ranch

% Moisture	84.1	% Phosphorous:	.40
% Protein		% Potassium:	4.00
Oven Dry Basis	21.2	% Calcium:	.86
As Fed Basis	3.4	% Magnesium:	.24
Estimated Digestible[1]	17.5	% Sulfur:	.54
% Acid Detergent Fiber:	23.6	ppm Sodium:	1625
		ppm Iron:	475
% TDN (Estimated):	--	ppm Manganese:	372
		ppm Zinc:	44
% NO_3^- (Nitrate):		ppm Copper:	11

[1]*Estimated Digestible Protein is calculated from Oven Dry Protein. A moisture of 10% is assumed.*

COMMENTS:

The fiber content (ADF) is too low to get an accurate TDN estimation.

Terry L. Bell
Associate Director
Soil Testing Laboratory

Important points to note in a basic forage analysis include crude protein level on a dry basis; a measure of digestibility, usually determined as Acid Detergent Fiber or Neutral Detergent Fiber; and calcium:phosphorus ratio. This cool-season combination of ryegrass and clover is producing well above the minimum of 14-16% protein. ADF value of less than 30 indicates the forage is low in fiber and has a high digestible volume. While calcium and phosphorus levels are adequately high, the ratio barely meets the required 2:1 ratio.

Chapter Six

Trouble-Shooting Food Plot Problems

When trouble-shooting food plot problems, it pays to take a close, systematic look. A windshield survey from your pickup can lead to false information. Get out in your food plot for a close inspection of whatever gremlins may be hiding there. Before correcting a problem, you must be sure you are managing for the right problem. That way you can avoid the frustration of taking action without getting a positive response. Trouble-shooting is a 2-stage process; first analyze and define the problem, then select possible solutions. Often there can be several solutions to a single problem, or there may be more than one problem. Let's take a look at some of the most common causes of food plot failure and discuss some of the probable solutions.

Causes of Crop Failure

1. Drought

Unfortunately, besides irrigation, we can do little if Mother Nature doesn't cooperate. No plant can grow without water. If you live in an area that receives less than 35 inches of rainfall annually, the chances of a crop failure from drought makes supplemental food plots risky at best, unless you are willing to invest in intensive dry land farming techniques. Even in areas that receive sufficient rainfall on an annual basis, inevitably there are times when rainfall is insufficient during the growing season to sustain your plantings. The result: partial or total crop failure.

If irrigation is possible, by all means use it. However, other options may offer at least a partial solution. Normally, spring is a wet period in the Southeast. Have your plans completed well in advance so there is no reason for delay when soil temperature and moisture are adequate for planting warm-season crops. Plants that are partially grown and have well-established root systems can deal with limited moisture more readily than young, tender seedlings.

In fall, if soil moisture is insufficient to germinate seeds, you can delay planting until enough rainfall is received. The downside to delayed planting is that your food plot may not be growing until well into the hunting season. However, it will be there for the stress period that occurs later in the winter. Unless you have an early freeze, an established warm-season planting usually will suffice until the cool-season plot is established. Combination plantings of cereal grains and legumes, again, are beneficial in cool-season plots because most cereal grains quickly germinate upon receiving moisture.

2. Improper Planting

A common problem is planting the wrong crop for the season. Remember, there are plants that grow during the summer or warm-season and plants that grow during the winter or cool-season, but usually not during both. We won't dwell upon this problem because if you followed the suggestions in this manual, it shouldn't occur. Another mistake is choosing a variety that is not adapted to your area or to the soil type where you want to establish the food plot.

Legumes that are properly inoculated form root nodules that "fix" atmospheric nitrogen into the soil. This nitrogen also can be used by associated plants and crops that follow.

If there is any question, consult your Agricultural Extension agent or reputable seed dealer about plant species and varieties that will grow successfully in your area.

However, as we noted earlier, the *most* common problem is improper planting technique. In particular, planting seeds at the wrong depth (see Chapter Four for detailed planting guides). Seeds need to be in firm contact with the soil. This way the tiny roots emerging from the seed can enter the soil immediately to begin growing and absorbing nutrients. Seeds planted too deep will die before breaking the soil surface in their search for sunlight. This can lead to a poor stand or even a complete failure of that plant species in your plot. It may seem as if you never planted them. The most common cause of failure in legumes is planting them too deep. Legumes with small seeds like clover and vetch never should be covered with more than 1/4 inch of soil.

Many species of legumes, like cowpeas, clovers and vetches, if properly inoculated, form root nodules that have the ability to fix atmospheric nitrogen into the soil. This supplemental nitrogen can increase natural soil fertility and lower fertilizer needs over time. However, for these nodules to form, an inoculant must be added to the seed at the time of planting. Failure to coat the seeds with the specific inoculant for that species will result in a poor stand that does not grow well.

Another problem that will end with the same result stems from allowing inoculated seed to become too hot. Inoculant is a "live" bacteria that can be killed by excessive heat. Inoculant should be stored in a

Over-utilization of forages on small food plots is a common problem, particularly with "ice cream" plants like cowpeas that deer will eat almost to the exclusion of other foods.

cool, dry place until you are ready to plant. A refrigerator is best. Mix only the amount of inoculant and seed necessary to fill your seeder and then broadcast or plant the mixture as soon as the sticking agent dries sufficiently for the seed to be dispersed. Also, while it would reduce your effort, do not mix and broadcast inoculated seed with fertilizer. Contact with strong fertilizer can kill the bacteria.

3. Excessive Utilization

Some plants, like cowpeas, often are referred to as "ice cream" plants, meaning they are preferred highly by deer and will be utilized almost to the exclusion of everything else. If you planted only peas, an entire crop may be wiped out in just a few days. Similar results can happen even with combination plantings if the acreage of your food plots is insufficient to handle the pressure of the deer herd. This benefits your deer only for a short time and defeats the purpose of establishing a supplemental food plot.

There are several possible solutions, all of which should be administered concurrently for a satisfactory resolution to the problem. Namely, control the deer population size, increase the amount of acreage planted, and use combination plantings of compatible species that can withstand and recover from heavy use. In most cases, utlization should not exceed 1/2 of the available forage.

4. Disease and Insects

More than 10,000 species of insects, 1500 plant diseases and 1500 species of nematodes are known to be harmful, in at least some degree, to humans, plants and animals. This makes determining the causative factor, and its ultimate control, a complex issue indeed. Fortunately, the common problems that you may have to deal with will not involve all of these.

A complete listing of possible diseases and their cures is far beyond the scope of this manual. However, some generalizations can be made. Most plant diseases are caused by fungus parasites, with a few others caused by bacteria and viruses. General symptoms to look for if you suspect a disease problem would be stunted plants, discolored leaf or stem tissue, various root and crown rots, or shrivelled and distorted leaves and seed heads.

Plants are attacked by numerous insects including aphids, grasshoppers, crickets, leafhoppers, army worms, cutworms, and an assorted array of caterpillars to name a few. Insects can attack plants both above and below ground. Ragged, chewed plants are a common characteristic. Above-ground insects will normally be easy to find if they are numerous enough to cause a serious problem in production. The greatest problem you may encounter with clovers is nematodes. Eventually, you may have to rotate crops to prevent continued problems. Some clovers, for example alyceclover, have nematode-resistant varieties available.

Monitor your crops and consult with a local professional should a disease or insect problem arise. Take a sample of the diseased plants or the insects you suspect to be causing the problem with you for identification. County Agricultural Extension agents usually are well trained in local pest problems and their solutions. Many times solutions can be as simple as rotating crops.

Spin-type feeders are commonly used to dispense grain to attract deer. However, they perform poorly when used as supplemental feeders.

Planning for a Crop Failure

No one ever "plans" for their crops to fail. But, if it happens, what options are available to keep supplemental nutrients available to your deer? Remember, if your crop failed because of drought, native plants that deer like to eat will be suffering as well. A good contingency plan, where legal, is to have a feeder available that can dispense pelleted feed. This is the same approach used to supplement deer in arid climates where food plots are risky at best. One free-choice feeder per food plot should be established.

Although commercially prepared feeds are marketed under various brand names, choose a pelleted feed that contains at least 16 but no more than 20 percent protein, a calcium to phosphorus ratio of 2:1 or better, and added essential vitamins, minerals and trace elements. Many pelleted feeds prepared for livestock will meet the dietary requirements of deer, although a number of manufacturers are marketing pelleted feeds specifically formulated for deer. Corn and "sweet feeds" like horse and mule chow are good sources of carbohydrates and fats, but do not contain enough

Commercial feeders are available that can be used to dispense pelleted supplemental feed for deer.

crude protein to meet the dietary requirements during the growing season. A high protein, high mineral ration should be made available in summer. Corn mixed with protein pellets is a good alternative to supplement deer during winter.

Spin feeders are good for dispensing grain to bait deer. However, their use in supplemental feeding should be limited. First and foremost, spin feeders usually do not deliver an adequate volume of feed. Second, spin feeders have problems dispensing pelleted feeds. Pellets become broken and any amount of moisture in the feed reservoir tends to clog the dispensing mechanism. Third, what feed is delivered is usually on a timed basis, determined by the operator, leaving the deer little choice as to when a supplement is consumed. Lastly, feed usually is dispensed upon the ground. Concentrated feeding from the soil surface enhances the probability of spreading diseases and introducing parasite infections.

Pelleted feeds are best delivered in an open, trough-type feeder that has a roof to protect the pellets from rain. Acceptable feeder designs are available through several commercial manufacturers. However, this type of feeder can be easily handmade. Easy access to the feed is critical, especially for bucks. Deer are reluctant to stick their heads into a closed box where they cannot see and might damage velvet-covered antlers. An open feeder with a roof at least 3 feet above the bottom of the trough should suffice.

Some maintenance of trough feeders is necessary. It is advisable to place drain holes in the bottom so water can escape if it does enter the feed box. Any contaminated or spoiled feed should be removed and replaced *immediately*. Also, disinfect the box with a weak solution of bleach and water if feed was allowed to mold or mildew.

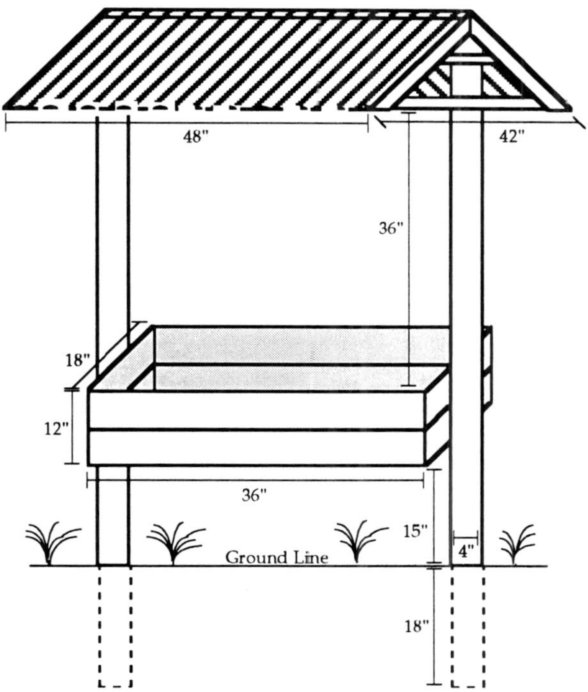

General plans for a covered deer feeder. Materials include Walmonized lumber and posts. Within the box, an 18 x 18 inch metal tray can be used to hold granular mineral. Trough dimensions are inside.

Chapter Seven

Natural Food Plots and Openings

There is no doubt establishing food plots can be expensive, especially if extensive clearing must be done. Expense alone can limit the acreage that can be established. However, native browse plants preferred by deer often occur in and near the edges of openings because of the increased amount of sunlight available. Heavy use by deer on native plants is common near planted plots because of the concentration of deer and the inherent increase in nutrition that results from fertilizer applications. These plants are valuable because they increase the forage available to deer above and beyond what you plant. Also, "weeds" in your food plot often provide just as much nutrition as the planted crop does! These same principles can be used to establish food plots that are

dominated by native plants, keeping cost and labor to a minimum while still providing high quality nutrition.

Natural Openings and Rights-of-Way

Natural meadows that are easily accessible and road, pipeline and utility line rights-of-way make excellent choices for natural food plots. The easiest way to apply nutrients is to drive a spreader truck down the road. Because fertilization increases the quality of plants, deer may actually seek out these areas that have been fertilized. You can create "hidden food plots" that are attractive to deer but are not easily recognized by other hunters.

"Natural" Foods

White-tailed deer are extremely adaptable creatures and, as such, consume a vast assortment of native plants. Specific preferences for particular plant species, however, vary by every region where whitetails are known to occur. Nevertheless, certain nutritional requirements and foraging strategies appear consistently throughout their range.

Deer forage basically can be divided into the following categories: 1) forbs (weeds); 2) grasses and grasslike plants; 3) mast (hard and soft fruits of woody plants); 4) mushrooms; and 5) browse (leaves and twigs of woody plants and vines). Deer will consume some of all of these categories. However, very few of the hundreds of plants occurring on deer range actually have significant value in their diet.

Over most of their range, deer prefer many species of forbs if available. Forb species selected by deer usually are highly nutritious and easily digested. However, forbs tend to be ephemeral and usually are found in abundance only during or shortly after rainy periods. After forbs mature, their nutritional quality and digestibility falls rapidly. Likewise, mushrooms can be important, particularly in phosphorus-poor soils, but occur primarily during the wetter portion of the year.

Mast, although low in protein, is a good source of carbohydrates necessary to meet the energy demands of deer during colder portions of the year. However, mast is highly variable in production and cannot be counted on no matter your management.

Native deer foods include forbs or weeds (upper left), grasses and grasslike plants (upper right), mast or soft and hard fruit of woody plants (middle left), mushrooms (middle right) and browse or leaves and twigs of woody plants and vines (lower left).

Grasses and grasslike plants make up only a very small percentage of the diet of deer, and then only during the early stages of plant growth. Certain grasses planted to "improved" pastures like bermudagrass have little value to deer.

Browse, on the other hand, is the dietary mainstay of most deer throughout their range. Browse is the most dependable portion of the plant community. Take the time to learn the browse plants preferred by deer in your area. Often these plants can easily be manipulated to increase their dietary value for deer. Fertilization of native browse not only can increase production but often increases palatability. Studies in east Texas found the growing season of browse plants could be extended by about a month with fertilization This adds another 30 days to the time the plants are actively growing and highly nutritious.

Additionally, some species readily adapt to transplanting if they do not occur where you would like. For example, honeysuckle, a highly preferred deer food where it occurs, has been transplanted successfully by cutting vines in the spring, spreading the cut vines over the transplant area, disking with a tractor into moist soil and fertilizing. Most native browse plants are perennial, meaning they come back from rootstock year after year, and could make excellent food plots if established and cared for. If you have a very high deer density, however, these plantings may be difficult to establish without fencing.

Fertilization and Liming of Native Plants

The association of wildlife abundance and soil fertility long has been recognized. Soil provides the nutrients to the plants that are ultimately passed on to the animals consuming the forage. Low soil fertility and/or an imbalance in soil nutrients results in low forage quality. This is true especially for phosphorus and nitrogen. A regular soil testing program determines exactly which nutrients are needed for optimum plant growth. A simple soil analysis can reduce or eliminate the addition of certain nutrients, thereby saving considerably in costs.

Most areas in the Southeast receive generous amounts of rainfall annually. However, abundant rainfall is a two-edged sword. Rainfall is necessary for optimum plant production, but, on the other hand, also quickly leaches nutrients from the soil. This results in an imbalance of soil nutrients and soil pH that is too acidic. Most forages do best in soils that are close to neutral.

To counteract these effects, soil samples should be tested annually to determine lime and fertilizer requirements for proper plant growth. Fertilizer and lime should be applied prior to spring green-up to obtain maximum benefits at rates determined from the soil test. This should never be the case but, if soil testing simply is impossible, a rule of thumb is natural food plots in the Southeast should be: 1) limed with agricultural limestone every 3 years at a rate of 2 tons per acre, 2) fertilized every year with 200 pounds per acre of 80-40-40, and 3) top-dressed with 100 pounds per acre of ammonium nitrate (34-0-0) at 60-day intervals or until plant growth ceases in the fall.

Be careful to locate natural food plots away from high runoff areas to keep fertilizer and lime from being washed into waterways. Also, as in planted food plots, natural food plots should be located near habitats frequented by deer.

Maintenance

Many native browse plants can grow quickly beyond the reach of deer. Any browse above about 4 feet in height is virtually unusable by deer as food. Resprouting species

sometimes need to be cut or pruned. Top-removal helps maintain a growth form that stays within the reach of deer. Because plants are the most nutritious when they are actively growing, annual fertilization is used to enhance the production and nutrient quality of sprouts and twigs.

The production of native forages for deer in the South is inversely related to timber density, primarily canopy closure. As canopy cover of overstory trees increases, the amount of sunlight able to reach the ground decreases. Deer forage is more effectively grown in open areas or at least under a sparse canopy.

Chapter Eight

Hunting Strategies for Food Plots

After two decades of deer research, we have come to the conclusion food plots are excellent for harvesting does, but seldom will a mature buck be taken from them. The lone exception is lightly hunted areas, where harvest is so low that bucks do not alter their movements and activity patterns in response to hunter pressure and hunter activity. However, using the right strategies, it is possible to significantly improve your chances of harvesting mature bucks while "hunting" your food plots.

Our research has often demonstrated that mature bucks quickly learn to pattern hunters. Probably the most serious mistake made by hunters in the South is using the same stand or blind year after year. Within a single season, deer quickly learn to adapt to a hunter's presence. They use the same area, but change their activity to avoid times when the hunter is present; usually after dark or mid-day.

In addition, how and what you plant also can influence your chances of success. You may be altering deer activity without knowing it, simply by the way you are planting your plots. Let's examine strategies to mitigate these two hunting problems.

Hunting Food Plots

One decided enemy of adequate deer harvest is the permanent blind or stand. In the South, many elaborate contraptions are used for hunting deer. We even saw a wrecked Plymouth car body erected on four stout poles being used as a hunting blind. Presumably, the hunter would roll down the driver's side window to make a shot!

Box blinds offer silent access and comfort, but are often avoided by mature bucks.

We strongly discourage you from using permanent stands on food plots. We also discourage hunting every plot on your property. This gives deer no safe haven to escape hunting pressure. Instead, we suggest you either alternate hunting on plots or use no stands at all. But, doesn't that go against traditional hunting techniques? You bet it does! Unfortunately, many hunters arrive at techniques because they once were successful. A fellow kills a nice 8-pointer in 1975 from his elevated blind, then hunts it until the day he dies, all the time thinking another nice buck will present itself. Season after season, however, he kills yearling bucks that are naive to hunter behavior.

There are two things to consider when hunting a food plot. First, deer behavior primarily is controlled by his sense of smell. The general pattern is a deer will approach a food plot from downwind, remain stationary for several minutes, then cautiously enter the plot. The areas where deer tend to congregate outside a plot are called *staging areas.* They serve social, as well as safety functions.

In the vast majority of the South, there is a prevailing wind during the fall, usually from a southerly direction. The exception, of course, is when northers blow in. Hence, in

Portable blinds enable hunters to easily position themselves in staging areas and travel lanes where deer are unaware of their presence.

most cases, deer will approach a plot into the prevailing wind—meaning approaching from a northerly direction. If you place your stand on the southern border of a plot, because it is convenient, you will be smelled by the vast majority of deer coming to the plot. On the other hand, if you position your stand so that it is off to one side away from an entrance trail, the probability of being detected is much less.

Instead of using permanent stands, we suggest you either use portable stands or even no stand at all. Portable stands come in many types, and they all work provided they do not take too long to erect. We recommend a stand that can be carried into the area before daylight, erected in less than 10 minutes and is light; less than 35 pounds. Some of the tripod and ladder stands fit these criteria, as well as the popular "climbing" stands. Also, a lock-on stand is useful, combined with climbing steps.

We prefer no stand at all. Rather, we use a pair of large pruning shears to cut aromatic vegetation and erect a ground blind from these materials. An experienced hunter can easily make a comfortable ground blind in less than 10 minutes. Pines, cedars, waxmyrtle and other aromatic trees or shrubs make excellent ground blinds. The oils produced by these plants make an effective cover scent. If you are leasing land, however, be sure and not cut commercially important trees.

Many hunters, especially those who have hunted out of permanent stands for many years, are uncomfortable in a ground blind. They tend to want to see great distances. However, the deer are coming to the plot, right? Why do you need to see hundreds of yards?

Lastly, we have found bucks tend to avoid plots during daylight hours. This does not mean they are not moving about. Rather, they have developed a unique movement strategy which allows them to cover a large area and monitor what's going on reproductively with the females. Not only do we use natural ground blinds, we seldom erect them on the food plot! That's right, we do not hunt directly on the plot.

Bucks will move adjacent to plots so they can monitor the plot by smell. Consequently, buck travel corridors tend to be downwind from the plot, usually about 20 to 50 yards. They also spend a great deal of time in and around staging areas, leaving signpost rubs to affect the estrous cycle of does and suppress subordinate bucks. Strategically place your stand or blind so you can intercept a buck before he arrives at a plot or staging area.

Planting Strategies

The second key is *what* you plant also affects your hunting success. Although we often have said the modern hunter is more concerned with the welfare of deer, he also is concerned with achieving proper harvest. This is good, both for the deer and the deer hunter. Therefore, you need to consider what you plant, and where, to effectively meet the harvest goals.

Many hunting clubs and landowners, in their zeal to plant something for deer, tend to plant cool-season mixtures in every available location on the property. One manager we know last season planted every road, pipeline right-of-way and plot to the same mixture of wheat, oats and arrowleaf clover. He did such a good job, a deer only had to walk a few yards to feed. His hunters were

While road rights-of-way make excellent food plots, these areas should be planted to cool-season plants that mature late like clovers, so deer will concentrate feeding on hunted food plots that are planted in a high proportion of early-maturing cereal grains.

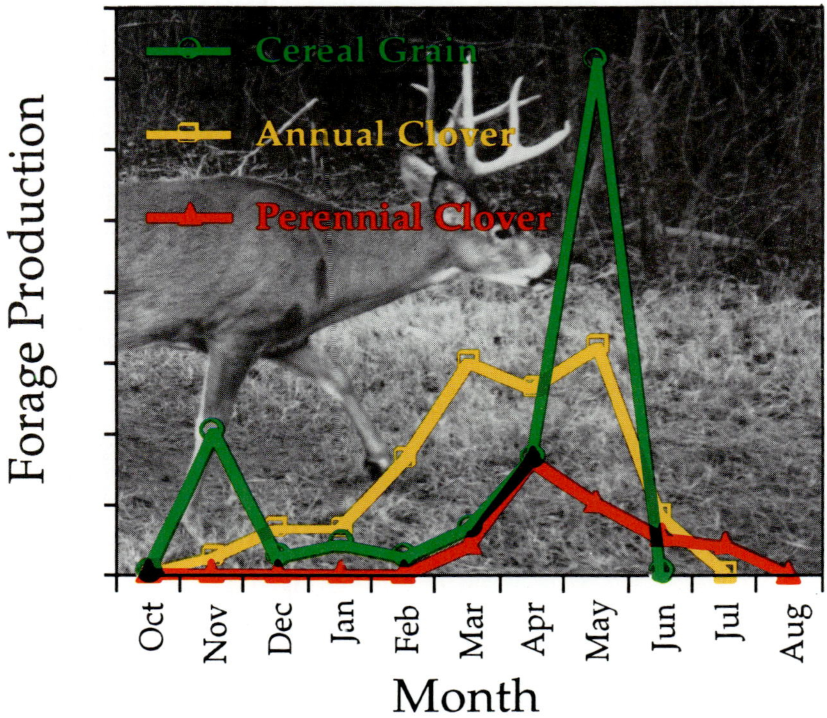

Cereal grains normally begin growth earlier than clovers. However, higher forage production can be realized with combination plantings that provide quality nutrition when deer need it the most.

---

very disappointed in the harvest, which fell much below previous years. He had spread his deer out too far.

Does this mean we should not make it easier for deer to feed? Not at all. Rather, you should consider altering your plantings to take advantage of deer movements. Studies in Mississippi by Dr. Harry Jacobson showed deer will not leave their established home ranges to feed on a food plot, but do alter movement within their home range to benefit from plots.

Long ago, fisheries biologists learned that to properly harvest fish from a large pond, the best approach was to draw down the water to a smaller pool. Obviously, you cannot do this with deer habitat,

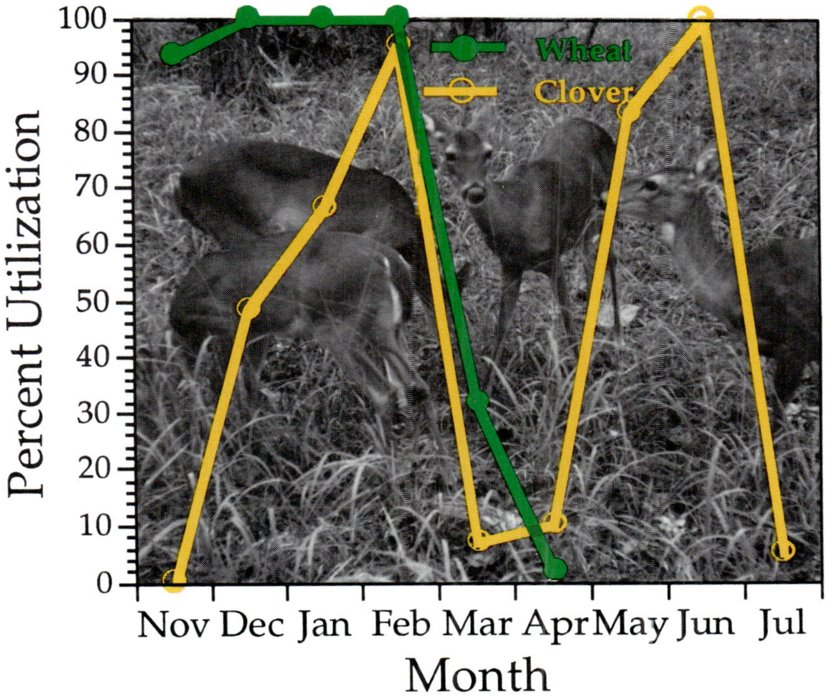

Deer take advantage of lush new growth of cereal grains early in the growing season and then switch to clovers as production of these plants increases in late winter.

but you can in a sense. Cereal grains and cool-season clovers have completely different growth strategies. The grains tend to put on significant growth in the cooler fall months; while legumes such as clovers establish themselves in fall, then accelerate growth beginning in late winter to early spring. In addition, cereal grains are highly digestible and nutritious early in the season. Consequently, deer feed more on the grains than clovers.

If you plant a mixture of cereal grains and clovers on every plot and right-of-way, you make them all equally attractive. But, if you limit plantings on non-hunted areas to legumes, concentrating cereal grains on hunted plots, you significantly alter deer movements. In essence, you have "drawn down your pond." The clovers planted on the non-hunted plots and rights-of-way

will be of great benefit when deer really need them in late winter before spring growth.

Finally, it can be beneficial to plant more expensive "ice cream" plants on hunted plots, as these varieties are highly attractive to deer. They should, however, be planted in combination with a more durable mixture, so the same plot will continue to benefit deer after the hunting season.

Chapter Nine

Economics of Food Plots

There always is a danger in dating a book by giving pricing information. To avoid this, we will discuss the economics of food plots in 1994 dollars. Prices continuously change and there is the added effect of inflation, so adjust cost estimates based on current rates. We provide this information primarily to 1) give you an idea of what activities cost, and 2) serve as a planning framework for your supplemental management program.

Site Preparation

Although we prefer to use no-till or low-till whenever possible, there is the need either to prepare the plot initially or to plant by traditional means. In either case, there is a real cost of site preparation. In general, you will encounter one of two conditions. Either there will be areas with little significant woody vegetation such as abandoned fields, rights-of-way or logging decks; or, you

Activity	Unit	Cost per Unit
Soil Analyses	Plot	$10.00
Fertilizer	Acre	$38.00
Liming	Acre	$25.00
Mowing	Acre	$15.00
Disking	Acre	$15.00
Seeding	Acre	$ 5.00
Total/Acre Site Preparation Cost		$108.00

Generalized cost for establishment of a supplemental food plot on a per acre basis.

will have to clear wooded areas prior to planting.

Obviously, the first scenario will involve the least cost. By taking advantage of existing openings you can substantially reduce your site preparation cost. In preparing your site, you should follow these steps:

1. *collect soil samples;*
2. *add soil amendments;*
3. *mow existing vegetation;*
4. *disk or harrow the site;*
5. *plant; and,*
6. *cover seed (unless using a drill).*

We generally budget as in the table above.

If, however, the land must be cleared you will have to add an additional $200 per acre for clearing cost. Afterwards, you still must prepare the site as given for previously cleared land. Therefore, depending on your initial situation, you should budget between $108 and $308 per acre for site preparation.

Seed

Notice we carefully left out the cost of seed. That depends on the combination or variety selected. As we have noted several times in this manual, it is important to match varieties to soil and site conditions. But, here are 2 common examples; one for warm-season and the other for cool-season.

We have found on upland soils, a combination planting

Activity	Unit	Cost per Unit
Soil Analyses	Plot	$10.00
Fertilizer	Acre	$38.00
Mowing	Acre	$15.00
Seeding	Acre	$ 5.00
Total/Acre Cost		$68.00

Generalized annual maintenance costs of supplemental food plots.

of iron and clay cowpeas and alyceclover will provide a summer-long, protein-rich crop. After experiments with many combinations, Dr. Billy Higginbotham found a mixture of 35-40 pounds of cowpeas and 15-20 pounds of alyceclover worked best. Total per acre cost for this planting is $49.50: cowpeas, $31.50 (35 lbs/ac.) and alyceclover, $18.00 (20 lbs/ac.). Inoculant for this combination will cost about $3.00, bringing total seed cost to $52.50 per acre. In moist, bottomland soils addition of 2-3 pounds per acre of American jointvetch will add about $6.00 to cost.

A cool-season combination planting of elbon rye, ryegrass and arrowleaf clover also will work well in upland plots. A per acre planting of 75 pounds elbon rye, 15 pounds ryegrass and 8 pounds arrowleaf clover will cost $26.70: rye, $15.00, ryegrass, $4.50 and arrowleaf clover, $7.20. Again, the clover seed must be inoculated at $3.00 per acre, totaling $29.70 per acre. In bottomland sites, arrowleaf clover does not do well so you will have to substitute the more expensive white clovers, but at a lower planting rate. These varieties usually cost about $1.50 to $2.00 per pound, and are planted at 4 pounds per acre, so total cost is about the same as for arrowleaf clover.

Total cost for a warm-season plot planted on a relatively clear site, therefore, is $160.50 per acre; while a cool-season planting will cost $137.70 per acre, initially.

Intermediate Treatments

We found just planting a cool-season crop is not enough. By late winter, soil nutrients such as nitrogen are depleted. Even though you have planted mixtures of clover and cereal grains, there often is not enough growth by the clover to adequately fix nitrogen during winter. As we noted earlier, we recommend supplemental applications of nitrogen for cool-season mixed plantings only. Normally, this will entail topdressing your plot with about 100 pounds per acre of a nitrogen source such as ammonium nitrate. You should include in your budget an additional $10.00 per acre for this activity.

Maintenance Costs

Once plots have been established, your per acre cost should decrease over time; provided you are doing a good job of plot management. Unless you have a very high deer population, most clovers will reseed annually. It makes no sense, therefore, to annually disk up this valuable seedbank. Rather, we recommend either mowing or light disking to preserve clover seed. Cereal grains such as elbon rye and wheat, as well as ryegrass can be topseeded or drilled through a thatch or existing crop. On the other hand, oats will not germinate adequately unless covered with soil, and do not fit well into a no-till program.

Unless you opt for annually replanting your plots as if from scratch, you should experience a significant cost reduction in subsequent years. Normally you will have to annually conduct soil analyses, fertilize according to soil test results, seed and mow your plots. If a sufficient clover seed bank exists in cool-season plots, you will only have to topseed or drill cereal grains and ryegrass. In summer plots, species such as alyceclover and jointvetch are excellent reseeders, while cowpeas and soybeans are not. Liming usually is required only every 3-5 years. We have found the annual maintenance costs shown above to be reasonable. Maintenance seed costs for the examples above would be warm-season, $31.50; cool-season, $19.50 per acre.

If you use only legumes in your plots, both warm and cool season, you can reduce your annual costs even more. We have some cool-season clover plots that have voluntarily reseeded for over 15 years! Annual maintenance costs for these plots involve only soil analyses, fertilizing and/or liming and mowing. We also discovered fertilizer costs diminish for clover plots, because there is no need for nitrogen; and, phosphorus and potash tend to remain in the soil for 2 or more years. In one field test, we were able to reduce fertilizer requirements to less than 50 pounds per acre phosphorus. *We would never have discovered this without an annual soil test.*

Summary

Studies by the Institute and Texas A&M University have suggested southern food plots are cost-effective. Average per acre yield of combination cool-season plantings of arrowleaf clover, rye and ryegrass may exceed seven tons! Using establishment costs only (the greatest expenditure), the per pound cost for a cool-season planting conservatively does not exceed $0.01 per pound.

plot preparation	**$108.00**
seed	**$ 29.70**
additional fertilizer	**<u>$ 10.00</u>**
Total	**$147.70**

$147.70 ÷ 14,000 pounds equals $0.01 per pound.

Warm-season plantings produce less than cool-season crops. Our east Texas average was about 2 tons per acre. Using the same basic cost analysis, the per pound cost is $0.04 per pound.

plot preparation	**$108.00**
seed	**$ 52.50**
Total	**$160.50**

$160.50 ÷ 4,000 pounds equals $.04 per pound.

Because a pelleted ration costs at least $0.25 per pound and corn, an inferior feed, costs around $0.10 per pound, we feel this certainly is cost-effective.

Earlier, we noted you should plant about 2 percent of your area to food plots; and, of this amount 60 percent should be in warm season and the remaining 40 percent in cool-season plots. As you probably noted, the lower expected yields on warm-season plots is the reason for the larger allocation to these plantings. Additionly, warm-season plantings have a greater impact on your herd's nutritional plane. Addition of food plots at this rate to your property adds $1.00 to $1.50 per acre to your management costs. If you are leasing, we suggest you pass this cost on to the lessee. Our experience is he is more than willing to pay the additional amount for *quality* food plots. Please note we did not include opportunity costs in our analysis. Opportunity costs are those associated with taking an acre of land out of production for other, possibly commercial, uses. Because land use varies, you will have to factor opportunity cost based on your particular situation. It is possible to reduce opportunity cost by utilizing areas, such as utility rights-of-way and permanent logging decks, not available for other uses.

Chapter Ten

Choosing Supplemental Forages

Supplemental forages have shown time and again in scientific studies to have positive effects on weight gain, antler development and productivity of deer herds if properly managed and the appropriate plant species used. Clovers and other legumes are the basic foundation for most successful food plots. These plants typically are highly nutritious, digestible and preferred by deer. However, contentions that one type or variety will produce significant results over another variety probably are unfounded. The most effective supplemental food plot contains nutritious species preferred by deer; and adapted to the specific soil and climatic conditions of the particular region in question.

Despite allegations to the contrary, no single species or even combination of species will be able to meet all of these criteria over a broad geographic range in all years. Many factors affect production from year to year, area to area, and even from plot to plot. Sometimes, large differences can be found even within the same plot that are difficult to explain. For this reason, make your selection based on species and varieties best suited to your area.

The following descriptions and basic managment are for plants commonly planted in food plots for deer in the Southeast. Generally, cool-season varieties are planted in the fall and do most of their growth during winter and early spring. Warm-season varieties are generally planted in spring to early summer and are intended to provide high quality forage during the warmer months.

Preferred Cool-Season Legumes

Alfalfa

Appearance and Growth: Upright to spreading perennial that produces numerous stems 2 to 3 ft tall.

Soil Adaptations: Best suited to well-drained, fine and medium textured soils. Especially intolerant of acid soils. Preferred soil pH is above 6.5.

Advantages or Disadvantages: Abundant forage production highly preferred by deer. Tolerance to drought is fair once established. Adapted to irrigation. Subject to cotton root rot. Annual soil tests are critical in maintaining high yields as levels of sulfur, potassium, phosphorus and boron determine production capabilities.

Management Considerations: Seeds must be inoculated prior to planting. Because of the need for medium to high

pH, this plant is not suited for many areas of the Southeast. Newly developed southern varieties show promise, but still require heavy lime applications. A firmly packed seedbed is required for till-planting. A low-till drill works if seeding into an established sod.

Arrowleaf Clover (Amclo, Meechee, Yuchi)

Appearance and Growth: An upright, cool-season annual clover that grows to a height of 40-50 inches under optimum conditions. A highly productive plant that normally attains peak growth in late winter and early spring. Blooming and seed maturation usually occurs in late June to early August.

Soil Adaptations: Suited to a wide range of soil conditions, but best adapted to fertile, well-drained sites. Arrowleaf clovers will not withstand calcareous or wet soil conditions. Preferred pH range is between 5.8 to 6.5.

Advantages or Disadvantages: Highly preferred by deer. As with most clovers, growth is typically slow after initial germination in the fall making it less attractive for deer during hunting season. However, abundant growth beginning in late winter provides excellent forage until early summer.

Management Considerations: Seeds must be inoculated prior to planting. Reseeds readily if adequate growth is allowed and not overgrazed. Contains a high proportion of hard seed. Mow in late summer after seed set, followed by light disking or late summer burning. May be necessary to lightly seed the plot each year until the stand is well developed. Applications of a low nitrogen fertilizer should be made after plants are established in the fall and followed-up with annual applications of low nitrogen fertilizer in late summer according to soil tests.

Crimson Clover

Appearance and Growth: An upright, cool-season annual that derives its name from the darkish red to scarlet-colored flowers. Crimson clover has soft hairy leaves and stems, and will reach a height of about 18 inches.

Soil Adaptations: Suited to a variety of sites but does best in upland clay soils that are well-drained. Tolerant of acid soils. Does not grow well in calcareous soils or wet sites. Preferred soil pH is between 5.5 and 7.0.

Advantages or Disadvantages: Preferred by deer. Reseeds readily and is often found growing abundantly in roadside ditches. Growth is slow in the fall but typically has forage available for grazing earlier than most clovers. Matures in late winter so forage is available only for a short period. Susceptible to cold damage.

Management Considerations: Seeds must be inoculated prior to planting. Apply low nitrogen fertilizer in the fall after the plants establish. Annual applications of low nitrogen fertilizer should be made in late summer according to soil tests.

Red Clover

Appearance and Growth: An erect cool-season legume that varies in growth form. It may act as an annual in the lower, warmer portions of the South or a short-lived perennial further north.

Soil Adaptations: More tolerant of soil acidity and poorly drained sites than alfalfa, but is less tolerant of wet sites than white clover. Maintain soil pH above 5.5.

Advantages or Disadvantages: Preferred by deer. Has a long growing season where well adapted allowing possibly the highest production of all clovers. Moderately drought tolerant but will not withstand heavy grazing.

Management Considerations: Seeds must be inoculated prior to planting. Apply low nitrogen fertilizer according to soil test recommendations in the fall after the plants are established. Usually will respond well to phosphorus and potassium fertilizer.

Subterranean Clover

Appearance and Growth: A cool-season annual with low-growing, non-rooting stems. Subterranean clovers are shade tolerant, but best growth occurs in open areas. Peak growth occurs in late winter to early spring.

Soil Adaptations: Best suited to well-drained sandy loam and clay upland sites. Does not grow well in wet soil conditions. More tolerant of low pH than most legumes. Preferred pH is between 5.5 and 7.0.

Advantages or Disadvantages: Highly preferred by deer. Growth is typically slow in the fall but provides abundant forage in late winter and early spring. Usually matures and becomes unavailable before the summer stress period. Seeds mature in a burr at or below groundline which allows for seed production

even under moderately intensive grazing. Reseeds less reliably than arrowleaf clover. The name is derived from its peanut-like habit of flowering and the developing seedhead growing into the ground for "self-planting."

Management Considerations: Seeds must be inoculated before planting. Apply a low nitrogen fertilizer after the plants become established. Mow or lightly disk in late summer and apply a low nitrogen fertilizer to help stimulate natural reseeding.

Vetch (Bigflower, Hairy)

Appearance and Growth: Vetches are upright, cool-season annuals with purple, white or yellow flowers. Semi-vining with weak stems and finely divided leaves.

Soil Adaptations: Best suited to moderate- and well-drained sites. Moderately sensitive to soil acidity and has a high phosphorus requirement.

Advantages or Disadvantages: Bigflower vetch is preferred by deer. Hairy vetch is less preferred. Little growth occurs in the fall but rapid development begins in late winter. Naturally reseeds if seed pods are allowed to mature. Forage production normally is less than clovers. Especially intolerant of grazing until about 6 inches tall. Grazing destroys the buds need for regrowth.

Management Considerations: Sufficient acreage should be planted so that plots are not overgrazed. Seeds must be inoculated prior to planting. Apply a low nitrogen fertilizer in the fall after the plants establish. Annual applications of low nitrogen fertilizer should be made in late summer or fall according to soil tests. Mow and lightly disk in late summer to stimulate natural reseeding.

White or Ladino Clover (numerous varieties)

Appearance and Growth: Depending on the variety, white or ladino clovers are either erect or low growing, cool-season perennials that often form extensive mats from creeping stems or runners. Runners have the ability to form roots at the joints or nodes, thereby allowing establishment of new plants. Leaves typically have a whitish v-shaped mark.

Soil Adaptations: Suited to a wide variety of soils including bottomland sites and other sub-irrigated soils. Preferred pH is between 6.0 to 6.5, but will tolerate moderately acid to moderately alkaline soils.

Advantages or Disadvantages: Highly preferred by deer. Initially slow growing, but will produce abundant quantity of excellent forage from late winter through early summer. Significant growth usually occurs again in early fall when cool night temperatures arrive. While leaves and leaf stems are highly palatable, jointed stems and runners tend to be tough and not easily eaten by deer. Being a perennial means it comes back from rootstock each year although new plants are established from seed and rooting ability of fallen stems. Not tolerant of drought.

Management Considerations: Seeds must be inoculated prior to planting. Apply low nitrogen fertilizer in the fall after the plants establish. Annual applications of low nitrogen fertilizer should be made in late summer according to soil tests. Usually highly responsive to potassium fertilizer.

Other Cool-Season Legumes

Alsike Clover

Appearance and Growth: An upright, short-lived cool-season perennial that can reach a height of 1 to 3 feet.

Soil Adaptations: Adapted to cool climates and poorly-drained soils. More tolerant of wet sites and low pH soils than most clovers.

Advantages or Disadvantages: Preferred by deer but will not tolerate drought conditions.

Management Considerations: Seeds must be inoculated prior to planting. Apply low nitrogen fertilizer in the fall after the plants establish. Annual applications of low nitrogen fertilizer should be made in late summer according to soil tests. Usually responds well to phosphorus and potassium.

Austrian Winter Pea

Appearance and Growth: Similar in growth habit and appearance to garden peas, except flowers and seed pods on Austrian winter peas are smaller. A vining plant that may reach a height of 2 feet.

Soil Adaptations: Does best on well-drained loam to sandy loam soils, but will grow in soils too wet for most clovers and small grains. Very sensitive to soil acidity. Prefers neutral to slightly alkaline soil pH.

Advantages or Disadvantages: A high yielding plant that is easily overgrazed. Excellent for low-till and no-till planting, especially in combination with cereal grains. Usually an early producer, but matures in early spring.

Management Considerations: Plant sufficient acreage to prevent overgrazing. Seeds must be inoculated prior to planting. Apply a low nitrogen fertilizer in the fall after the plants establish. Annual applications of low nitrogen fertilizer should be made in late summer according to soil tests. Mow and lightly disk in late summer to stimulate natural reseeding.

Berseem Clover

Appearance and Growth: An upright cool-season annual

that somewhat resembles alfalfa. Stems may reach a height of 2 feet or more.

Soil Adaptations: Tolerant of alkaline and poorly-drained soil. Performs best on loam soils with pH of 6.0 or above.

Advantages or Disadvantages: Preferred by deer but is not very tolerant of freezing conditions. Requires highly fertile soil with a high boron content. Varieties that are winter-hardy produce hard seed and will often reseed. Well suited for non-acid Black Belt soils and high rainfall areas near the Gulf Coast.

Management Considerations: Seeds must be inoculated prior to planting. Apply low nitrogen fertilizer in the fall after the plants establish. Annual applications of low nitrogen fertilizer should be made in late summer according to soil tests.

Birdsfoot Trefoil

Appearance and Growth: A prostrate or erect cool-season perennial. Seed pods radiate from the stem to resemble a bird's foot.

Soil Adaptations: Suited for most well-drained soils. Tolerant of moderate soil acidity. Preferred soil pH is above 5.5

Advantages or Disadvantages: High quality forage. Somewhat drought tolerant. Grows best in the northern regions of the Southeast. Tends to be susceptible to crown and root rot.

Management Considerations: Plant sufficient acreage to prevent overgrazing. Seeds must be inoculated prior to planting. Apply a low nitrogen fertilizer in the fall after the plants establish. Annual applications of low nitrogen fertilizer should be made in late summer according to soil tests. Responds well to phosphorus and potassium fertilizer.

Rose Clover

Appearance and Growth: A spreading, cool-season annual that may reach up to 18 inches in height.

Soil Adaptations: Suited to most well-drained soils. Tolerant of low soil fertility. Optimum pH range is 6.0 to 7.0.

Advantages or Disadvantages: Highly preferred by deer.

Tolerant of drought and can withstand moderate grazing. Typically reseeds naturally, but some varieties produce low yields.

Management Considerations: Sufficient acreage needs to be established to prevent deer from overgrazing. Seeds must be inoculated prior to planting. Apply a low nitrogen fertilizer in the fall after the plants establish. Annual applications of low nitrogen fertilizer should be made in late summer according to soil tests. Usually responds well to phosphorus and potassium fertilizer.

Singletary Pea

Appearance and Growth: Upright, cool-season annual that forms vines up to 3 feet in length.

Soil Adaptations: Does best on well-drained sites but will grow in soils too wet for clovers and cereal grains. Very sensitive to soil acidity. Prefers neutral to slightly alkaline soil pH.

Advantages or Disadvantages: Highly preferred by deer. Can be easily overgrazed. Excellent for low-till or no-till planting.

Management Considerations: Plant sufficient acreage to prevent overgrazing. Seeds must be inoculated prior to planting. Apply a low nitrogen fertilizer in the fall after the plants establish. Annual applications of low nitrogen fertilizer should be made in late summer according to soil test. Mow and lightly disk in late summer.

Yellow or White Sweetclover

Appearance and Growth: Bushy, cool-season biennial legume that may persist well into summer. Stems may reach a height of up to 8 feet under ideal conditions.

Soil Adaptations: Best suited to well-drained clay or loam soils. Prefers neutral to alkaline soils with optimum pH between 6.5 to 7.5.

Advantages or Disadvantages: Preferred by deer. Will tolerate drought conditions and more cold tolerant than alfalfa. Will begin growth early in winter under mild conditions in the Southeast. Grows well in partial shade or in openings but is intolerant of dense shade. A biennial means it will take 2 years to produce seed, after which the plants die out and persistence of the stand depends upon natural or hand reseeding. Normally reseeds well in humid regions. Moderately tolerant of heavy grazing.

Management Considerations: Seeds must be inoculated prior to planting. Apply low nitrogen fertilizer according to soil test recommendations in the fall after the plants are established. Look for a low-coumarin variety as coumarin reduces palatability of the forage.

Cool-Season Cereal Grains

Oats

Appearance and Growth: An upright, cool-season annual with a grass-like appearance. Although a hard-coated seed is produced at maturity, leaf growth is the primary concern as supplemental feed.

Soil Adaptations: Suited to most well-drained soils except deep sands. Does not do well on poorly-drained sites.

Advantages or Disadvantages: Highly preferred by deer. Oats are easily established and usually begin growth in early fall making it highly attractive to deer during the hunting season. Must be reseeded each year. Normally not as tolerant of cold as wheat or rye. Cold-hardy varieties like Buck Magnet Oats should be considered. Oats are not suitable for no-till planting. Seeds must be covered, or at least in firm contact, with mineral soil for germination.

Management Considerations: Oats require little attention once planted. Should be fertilized at the time of

planting according to soil tests. Followup fertilization about 60 days after planting is recommended to achieve best production.

Rye
(several varieties)

Appearance and Growth: An upright, cool-season annual similar in appearance to wheat. Stems have a bluish appearance and may reach up to 4 feet in height.

Soil Adaptations: Suited to well-drained soils including sandy sites. Does not tolerate wet sites as well as wheat. Probably the most drought hardy of all cereal grains.

Advantages or Disadvantages: Highly preferred by deer. More tolerant of cold than either oats or wheat. Will reseed if allowed to mature but to ensure a good stand it should be seeded each year. Excellent for no-till planting.

Management Considerations: Rye requires little attention once planted. Should be fertilized at the time of planting following soil test recommendations. Followup fertilization by top-dressing about 60 days after planting is necessary to achieve best production.

Wheat

Appearance and Growth: An upright, cool-season annual with a grass-like appearance. Although a hard-coated seed is produced at maturity, the leaf growth is the primary concern as supplemental feed.

Soil Adaptations: Suited for most well-drained soils to moist soils.

Advantages or Disadvantages: Highly preferred by deer. Wheat is easily established and begins growth early in the fall making it highly attractive to deer during the hunting season. Tolerant of cold. Does not reseed well naturally. Must be reseeded each year. If left standing in the spring, the seed provides excellent food for quail and turkey. Can be low-tilled or no-tilled easily.

Management Considerations: Requires little attention once planted. Should be fertilized at the time of planting according to soil tests. All cool-season small grains are highly responsive to nitrogen. Followup fertilization by top-dressing about 60 days after planting is necessary to achieve best production.

Cool-Season Grass

Ryegrass

Appearance and Growth: An upright, cool-season bunch grass with a flattened seedhead.

Soil Adaptations: Suited to most moderate to well-drained sites. Will tolerate wetter soil conditions than oats.

Advantages or Disadvantages: Moderately preferred by deer. Will establish easily under most conditions. Will not provide growth as early as oats but continues to produce forage later into winter. Reseeds naturally if allowed to mature, although to ensure a good stand it should be seeded each year. More tolerant of cold than oats.

Management Considerations: Little attention is required once planted. Should be fertilized at the time of

planting following soil tests. Fertilizing by top-dressing at 60 days after planting is necessary to get maximum production. Mowing and lightly disking in late summer serves to stimulate natural reseeding.

Preferred Warm-Season Legumes

Alyceclover (Summer sweet clover)

Appearance and Growth: A warm-season annual with oval, light green leaves that are borne along the entire length of slender stems. Can reach a height of up to 3 feet in good conditions. Grows upright in dense stands but low and spreading in sparse stands or when the tops are removed by grazing. Despite its name this is not a true clover.

Soil Adaptations: Best suited for most moderate to well-drained sandy sites. Does not grow well on wet sites. Somewhat tolerant to acid soils.

Advantages or Disadvantages: Highly preferred by deer. Like American jointvetch, seedlings are not very competitive. Will produce abundant forage from late spring until the first frost. Grazing causes the stems to branch and creates a low growth form. Very tolerant of dry conditions once established. Can easily be planted with no-till methods.

Management Considerations: Seeds must be inoculated prior to planting. Fertilize with a high phosphorus and potash but low nitrogen fertilizer at seeding. Do not use nitrogen fertilizer.

American Jointvetch (Aeschynomene)

Appearance and Growth: An upright, warm-season annual with finely divided leaves

that reaches a height of over 3 feet.

Soil Adaptations: Best suited for rich sandy loam to silt loam soils on a slight slope including alluvial soils that are well-drained. Very dry conditions will not produce a good stand. Moderately sensitive to soil pH.

Advantages or Disadvantages: Highly preferred by deer. Vigorously reseeds if seed pods are allowed to mature. However, seedlings do not compete aggressively and native grass and weed competition can limit stand establishment. Seeding by hand is not recommended because throwing tends to bunch the seeds causing competition. Can be no-till planted using a seed spreader. Once established, abundant forage is produced from late spring until frost. Can withstand moderate to heavy grazing.

Management Considerations: Seeds must be inoculated prior to planting. After seeding, apply low nitrogen fertilizer after seedlings are established. In established stands, lightly disk and top-dress with a low nitrogen fertilizer according to soil tests in early spring.

Cowpeas
(Numerous varieties)

Appearance and Growth: Erect, warm-season annuals that may sometimes vine on weak stems.

Soil Adaptations: Widely

adapted to most soil types except deep sands and wet areas. Tolerant of soil acidity.

Advantages or Disadvantages: High forage production in early spring that is highly preferred by deer. Utilization at times is excessive because of the high palatability of these plants. Does not tolerate heavy grazing pressure. Grows well in combination with warm-season clovers by taking grazing pressure away from clover seedlings until they are well established. Drought tolerance is fair.

Management Considerations: Seeds must be inoculated prior to planting. Apply low nitrogen fertilizer according to soil test recommendations at establishment. Must be seeded each year. A large area must be planted to prevent overgrazing early.

Hairy Indigo

Appearance and Growth: An erect, warm-season legume that forms a deep taproot. Easily reaches up to 3 feet in height.

Soil Adaptations: Best suited for well-drained upland sites including sandy soils.

Advantages or Disadvantages: Abundant growth in late summer and fall. Because the deep taproot can reach to water sources largely unavailable to other plants, drought tolerance is very good. May not be grazed extensively if other forage is available. However, hairy indigo remains available when other plants have succumbed to lack of water. Extremely drought-hardy characteristics makes it a good "insurance" plant if combined with species of higher preference by deer.

Management Considerations: Seeds must be inoculated prior to planting. Apply low nitrogen fertilizer according to soil test recommendations at time of establishment.

Lablab

Appearance and Growth: An erect, warm-season legume that is a weak perennial.

Soil Adaptations: Preferred sites are well-drained sandy uplands in moist climates. Sandy loam to clay loam in more arid regions. Will not tolerate wet soil conditions. Best growth occurs in soils with pH between 5 and 7.5.

Advantages or Disadvantages: Highly preferred by deer. Does not readily reseed but produces abundant forage through the warmer months until first frost. Very drought tolerant. Seedlings are not very competitive so stands must kept weed-free. Must be protected from grazing until about the first month following establishment. Thereafter, it is tolerant of moderate to heavy grazing pressure.

Management Considerations: Seeds must be inoculated prior to planting. Apply low nitrogen fertilizer according to soil test recommendations at establishment. Does not reseed well naturally; must be seeded each year. May be best adapted warm-season legume for arid regions like southern Texas, but usually will not produce as well as pea and clover varieties in higher rainfall areas. Intensive dry land farming must be used in arid regions.

Partridge Pea

Appearance and Growth: An upright, annual legume that can reach a height of up to 2 ft. Showy, yellow flowers are clustered in the leaf axils.

Soil Adaptations: Best suited to low, poorly-drained sandy to clay soils. Somewhat tolerant of acid soils but best growth occurs in soils with pH between 6 0 and 6.5.

Advantages or Disadvantages: Although the foliage is eaten somewhat by deer, the

majority of species known to use partridge pea are upland game birds. The hard, black seeds can be an important food source for bobwhites in fall and early winter. Very susceptible to drought and will not withstand heavy grazing. Will successfully reseed for several years under favorable conditions.

Management Considerations: Seeds must be inoculated prior to planting. Apply low nitrogen fertilizer according to soil test recommendations at time of establishment. Light disking in spring after the danger of frost is passed will improve establishment of volunteer seedlings.

Peanuts
(Numerous varieties)

Appearance and Growth: An upright, leafy perennial that reaches 12 to 16 inches.

Soil Adaptations: Grows best on well-drained sandy soil. Will not tolerate poorly-drained sites or acid soils.

Advantages or Disadvantages: Must be planted with rhizomes. Perennial peanut produces few seeds. Well adapted to the lower regions of the South. May require up to 2 years to establish a productive stand.

Management Considerations: Must be propagated from rhizomes. Should not be overgrazed before the first killing frost to allow root reserves to be replaced.

Soybean
(Several varieties)

Appearance and Growth: Erect, warm-season annuals that sometimes trail or vine. Typically a bushy plant with many leaves.

Soil Adaptations: Preferred sites are well-drained soils with pH of 5.8 or greater.

Advantages or Disadvantages: Highly preferred by deer, especially the less hairy-stemmed varieties. Produces a hard seed useful for dove, quail, and turkey. Produces abundant forage throughout the summer under good growing conditions but is somewhat sensitive to drought and does not tolerate heavy grazing pressure.

Management Considerations: Seeds must be inoculated prior to planting. Fertilize with a high phosphorus and potash but low nitrogen fertilizer at seeding. Can be broadcast, row-cropped or drilled, but germination is best if lightly covered with soil.

Other Warm-Season Legumes

Annual Lespedezas
(Numerous varieties)

Appearance and Growth: An erect, warm-season annual with fine stems and shallow taproots.

Soil Adaptations: Widely adapted except for deep sands or on soils that are poorly-drained. Preferred soil pH is between 5.0 and 6.5.

Advantages or Disadvantages: Vigorously reseeds. Seeds are choice food for northern bobwhites, and to some extent, for turkey and dove. Annual lespedezas are used sparingly by deer. Drought tolerance is poor to fair. On soils with high fertility, lespedezas may be crowded out by more productive grasses and clovers.

Management Considerations: Seeds must be inoculated prior to planting. Apply low nitrogen fertilizer according to soil test recommendations at the time of establishment. Mowing and lightly disking or controlled burning in fall helps reestablish the stand.

Bicolor Lespedeza

Appearance and Growth: Upright, perennial shrub that grows from 3 to 10 ft tall. The name "bicolor" comes from the flowers being two-colored; white and purple.

Soil Adaptations: Widely adapted except for deep sands or on sites that are poorly-drained. Soil pH should be maintained between 5.0 and 6.5.

Advantages or Disadvantages: Preferred by deer, however does not stand up to heavy browsing in the growing season. This imported legume can be important cover and food for quail and turkey if deer are not allowed to over-browse. If planted for deer, be sure you do not plant a "browse proof" variety as this will result in little utilization.

Management Considerations: Bicolor is a perennial and can be maintained indefinitely once established and if properly cared for. Can be planted from seed or by transplanting seedlings. Seeds must be inoculated prior to planting. Apply low nitrogen fertilizer according to soil test recommendations at the time of establishment. It's necessary to cut stems off near the groundline every three to five years during February to remove dead stems and rejuvenate the stand. Fertilize with a low nitrogen fertilizer each time the stand is pruned.

Velvet Beans

Appearance and Growth: A viney, annual legume that may extend to 40 feet.

Soil Adaptations: Adapted to sandy soils. Tolerant of soil acidity and low fertility.

Advantages or Disadvantages: Not highly preferred by deer. Hairs on the seed pods is a skin irritant to humans. Does not withstand heavy grazing.

Management Considerations: Often planted with a low seeding rate of corn as the stalks support the long, trailing vines. Seeds must be inoculated prior to planting. Apply low nitrogen fertilizer according to soil tests at time of establishment.

Warm-Season Grasses

Japanese Millet

Appearance and Growth: An erect, warm-season annual. Growth of this bunch grass can attain a height of one to eight feet depending on variety, site and fertility.

Soil Adaptations: Adapted to wet-site conditions. In high precipitation areas it can be grown on well-drained, upland soils. Preferred soil pH is between 5.4 and 7.4.

Advantages or Disadvantages: Abundant leafy growth that is moderately preferred by deer. Produces a hard seed that is particularly useful for waterfowl, and to a lesser extent, for doves, quail, and turkeys. Voluntarily reseeds well if proper conditions are provided. Drought tolerance is very low except in areas that can be flooded or irrigated. May be used as a nurse plant for slower growing legumes.

Management Considerations: Apply fertilizer according to soil tests. Volunteer seeding is improved by scarifying the soil between time of seed maturity and the following germinating season on dry sites, or drain the site and maintain a low water level until after germination on flooded wetlands.

Other Millets (Browntop, Common, Dove Proso, German, Pearl)

Appearance and Growth: Erect, warm-season, annual bunch grasses.

Soil Adaptations: Suited for

well-drained uplands.

Advantages or Disadvantages: Abundant forage that is used sparingly by deer. Seeds can be an important food source for dove, quail, and turkey. Also for waterfowl if planted near water. High moisture requirement results in drought tolerance being fair to poor, with Proso millet the most tolerant millet to dry conditions. May be useful as a nurse crop for slower growing legumes planted in combination.

Management Considerations: Apply fertilizer according to soil test recommendations. For the best results, the stand will have to be seeded each year.

Bahiagrass

Appearance and Growth: An

erect, warm-season perennial with a branched seedhead. Bahiagrass often forms a permanent sod.

Soil Adaptations: Widely adapted except for wet sites.

Advantages or Disadvantages: Forage is grazed sparingly by deer and then only when in very young growth stages. Drought tolerance is good. The major benefits are easy establishment, soil holding ability and favorable growth form that greatly enhances no-till planting.

Management Considerations: Can be planted in spring or fall, but best stands are usually established with fall planting. Established stands need little maintenance. May "fix" nitrogen similar to legumes to some extent.

Corn

Appearance and Growth: An erect, warm-season annual with thick stems.

Soil Adaptations: Best suited to well-drained upland sites, but will grow in bottoms as well.

Advantages or Disadvantages: Requires large acreages on highly fertile soil to prevent overgrazing and provides little high protein forage. However, standing corn can be left in the field for winter food. Not suited for no-till planting.

Management Considerations: Should be cultivated and maintained weed-free to produce best crops. Plant in early spring and top-dress with supplemental nitrogen fertilizer at 28-day intervals. Cannot be no-tilled.

Egyptian Wheat

Appearance and Growth: An erect, warm-season annual sorghum that can produce slender stalks from 7 to 10 feet in height.

Soil Adaptations: Suited to well-drained sites with full or lightly-filtered sunlight. Does poorly in deep sands.

Advantages or Disadvantages: Primary use is supplemental grain for quail. Seeds mature late and benefit quail in late fall and winter. Foliage usually is not browsed by deer and seed may be used only slightly. Spindly seed heads prevent excessive use by blackbirds and other perching birds.

Management Considerations: Should be row-cropped or drilled; broadcast plantings are usually unsatisfactory. Fertilize according to soil test recommendations. For best results, the stand will have to be seeded every year.

Grain Sorghum (W.G.F. and others)

Appearance and Growth: An erect, warm-season annual that is somewhat shorter in stature than most grain sorghum varieties. WGF stands for Wild Game Food.

Soil Adaptations: Widely adapted to well-drained upland sites. Very sensitive to soil acidity. Maintain pH above 5.6.

Advantages or Disadvantages: Produces an abundance of hard-shelled seed that resist mold and rotting. Seeds have a tannin-like substance that gives the grain a bitter taste until mature. While deer will consume the forage and some grain, this plant is usually established primarily for game birds like dove, quail, and turkey. Tolerance to drought is good. Nitrate accumulation, a highly toxic compound, can form in the plant when stressed by heat and drought.

Management Considerations: Apply fertilizer according to soil test recommendations. Does not reseed well so the stand will have to be seeded each year. Heavy grazing is required to maintain high quality in the rapidly growing stems.

Other Warm-Season Plants

Buckwheat

Appearance and Growth: A slow growing plant with an extremely short growing season.

Soil Adaptations: Wet soils and well-drained, rich, loamy soils.

Advantages or Disadvantages: Can be over-grazed easily, but is highly preferred by deer. Large areas must be planted to prevent deer from overgrazing. Can be planted several times during the

growing season. Provides excellent food for waterfowl as well.

Management Considerations: Can be broadcast seeded. Fertilize according to soil test recommendations.

Chufa

Appearance and Growth: An erect, warm-season perennial with a grass-like appearance.

Soil Adaptations: Best suited to well-drained sandy to sandy loam sites. Will grow in clay soils but the tuber will be difficult for turkeys to scratch out. Preferred pH range is from 5.0 to 7.5.

Advantages or Disadvantages: Moderately preferred by deer. Chufa produces an underground tuber that is similar to a peanut. While sometimes planted for deer, chufa is the primary supplemental food crop planted for wild turkeys. Also, especially favored by feral hogs. Not tolerant of heavy grazing. Expensive to plant and maintain. Does not reseed easily.

Management Considerations: Can be planted either row crop or broadcast. Fertilize according to soil tests at the time of planting and annually in the spring. Particular attention should be paid to potassium.

Great or Common Sunflower

Appearance and Growth: An erect, warm-season annual that can reach a height of six feet or more with showy yellow flowers.

Soil Adaptations: Widely adapted to most sites that are well-drained.

Advantages or Disadvantages: Used sparingly if at all by deer, and then primarily only in early growth stages. Sunflowers produce a high fat and carbohydrate seed that is used extensively by dove and quail, and to a lesser extent, by turkey. Wild varieties vigorously reseed. Has high drought tolerance. Large flocks of songbirds, especially blackbirds, are commonly attracted to sunflowers and can cause serious reductions in seed production.

Management Considerations: Wild, annual sunflowers can easily escape cultivation, causing "weed" problems where it is not wanted. Disking in late fall or winter rejuvenates the stand. Domestic varieties are not as prone to cause "weed" problems but are highly susceptible to insect damage. Domestic varieties usually must be reseeded each year.

Dwarf Essex Rape

Appearance and Growth: An erect, warm-season perennial that resembles turnips.

Soil Adaptations: Suited for damp soils.

Advantages or Disadvantages: A high protein content plant. Deer will browse the tops and later may consume to the turnip-like roots.

Management Considerations: A good plant to include in no-till seed combinations. Fertilize and lime according to soil tests at time of establishment.

Summary

Plant species and varieties described in this chapter are proven producers that have been used successfully in planting trials by agronomists and in management situations throughout the Southeast. One other common denominator is most of these species can be obtained readily by local seed distributors. However, this is not meant to imply these are the only plants that will work. If you find a plant that grows well in your area and can meet all of the criteria given at the beginning of this chapter for successful food plots; by all means use it.

General guidelines for seeding are given in Table 1 at the end of this chapter. In addition to supplying a large variety of seed, suppliers of specialty seeds, seed mixtures and native wildflower seeds have been listed in Appendix I at the end of this manual.

Table 1. Planting chart for supplemental food plots.

Crop Species	Seeding Rate(lbs/ac)*	Planting Date	Planting Depth
Cool-Season Legumes			
Arrowleaf clover	5-10	Sep-Nov	1/4-1/2"
Subterranean clover	10-20	Sep-Nov	1/4-1/2"
Crimson clover	20-25	Sep-Nov	1/4-1/2"
White clover	3-6	Sep-Nov	1/4-1/2"
Alsike clover	4-6	Sep-Nov	1/4-1/2"
Sweetclover	10-15	Sep-Nov	1/4-1/2"
Red clover	12-15	Sep-Oct	1/4-1/2"
Berseem clover	15-20	Sep-Oct	1/4-1/2"
Alfalfa	15-20	Sep-Oct	1/4-1/2"
Vetch	20-25	Sep-Oct	1/4-1/2"
Singletary pea	25-35	Sep-Nov	1-2"
Austrian winter pea	25-35	Sep-Nov	1-2"
Rose clover	15-20	Sep-Oct	1/4-1/2"
Birdsfoot trefoil	4-6	Aug-Sep	1/4-1/2"
Cool-Season Cereal Grains			
Wheat	75-120	Sep-Nov	1/2-1"
Oats	50-100	Sep-Nov	1/2-1"
Rye	84-112	Sep-Nov	1/2-1"
Cool-Season Grass			
Ryegrass	25-35	Sep-Nov	1/4-1/2"
Warm-Season Legumes			
Amerian jointvetch	5-10	Mar-Jun	1/2-1"
Alyceclover	10-20	Mar-Jun	1/4-1/2"
Soybean	60-75	Apr-Jun	1-2"
Cowpeas	60-90	Apr-Jun	1/2-1"
Lablab		Apr-Jun	
Partridge pea	5-10	Feb-Jun	1/2-1"
Velvet beans	30-35	May-Jun	3-4"
Bicolor lespedeza	10	Mar-May	1/2-1"
Annual lespedezas	25-30	Mar-May	1/2-1"

Table 1. Continued

Crop Species	Seeding Rate(lbs/ac)*	Planting Date	Planting Depth
Peanut	60-80 (bushels/ac. of rhizomes)	Dec-Mar	
Hairy Indigo	10-20	May-Jun	1-1 1/2
Warm-Season Grasses			
Japanese millet	30-35	Apr-Jul	1/2-1"
Other millets	35-50	Apr-Jul	1/2-1"
Bahiagrass	15-20	Mar-Jun	1/4-1/2"
Corn	12-20	Mar-May	1/2-1"
Grain sorghum	15-20	May-Jun	1-1 1/2"
Egyptian wheat	15-20	Apr-Jun	1-1 1/2"
Other Warm-Season Plants			
Buckwheat	50	May	1/4-1/2"
Chufa	25-50	May-Jun	1/2-1"
Rape	10-12	Apr-Jul	1/2-1"
Sunflower	10-30	Apr-May	1-2"

* Seeding rate is for planting a single species. Rate should be reduced by half when planted in combination with another single species. Reduce rate by 2/3 when planting with 2 or more species.

Appendix I

Wildlife and Specialty Seed Suppliers

At times beginners and experienced people as well, after reading a manual such as this, are inclined to try something new, but lack information on where and how to find the services. That is the reason for the lists provided in Appendices I and II. These lists are not intended to be a comprehensive coverage of all companies or products, but to provide a starting place. Listing of brand names, individuals or companies in this publication is solely to provide a reference and is not implied or intended to constitute an endorsement or recommendation by the authors, the Institute for White-tailed Deer Management and Research or by Stephen F. Austin State University.

BUCK MAGNET Forage Oats and Soybeans

Arkansas County Seed Company, Inc.
PO Box 43
Stuttgart, AR 72160
(501) 673-2706

Distributors of BUCK MAGNET seeds

Southern Farmers Services
PO Box 5489
N. Little Rock, AR 72119
(501) 945-2371

Trophy Hunting Products, Inc.
Rt. 3 Box 274
Lonoke, AR 72086
(501) 676-7288

Alabama Farmers Coop
PO Box 2227
Decatur, AL 35602
(205) 353-6843

Athens Seed Co.
PO Box 357
Watkinsville, GA 30677
(800) 282-7004

Merschman Seed, Inc.
103 Avenue D Box 67
West Point, IA 52656
(319) 837-6111

Growmark, Inc.
2205 Brush College
Decatur, IL 62526
(217) 423-9738

Ten Barge Seed Co.
PO Box 187
Haubstadt, IN 47639
(812) 768-6157

Sugar Creek Outfitters
6866 West 399 St
LaCygne, KS 66040
(800) 654-7225

Missouri Southern Seed
Box 699
Rolla, MO 65401
(314) 364-1336

Delta Purchasing
Federation
Box 8177
Greenwood, MS 38930
(601) 453-7374

Belleplain Supply
346 Handsmill Rd
Belleplain, NJ 08270
(609) 861-2345

Madison Seed Co.
13455 State Rd. 38 SE
London, OH 43140
(614) 874-9410

Ruff's Seed Farms
3870 Ruff Rd
Amanda, OH 43102
(614) 969-2600

Tri-Star Seed
PO Box 27186
Knoxville, TN 37927
(615) 522-8659

W. H. Anton Seed Co.
PO Box 667
Lockhart, TX 78644
(512) 398-2433

MBS Seed, Inc.
2534 Craig
PO Box 2939
Denton, TX 76202
(817) 387-2701

M-G Farm Service
PO Box 697
Weimer, TX 78962
(409) 725-8584

ESCO Distributors
404 N. Main
Box 22
Schulenburg, TX 78956
(409) 743-4141

Douglas W. King Co.
PO Box 200320
San Antonio, TX
(210) 661-4191
(210) 661-4191

Abilene Ag Service
1301 South Treadway
Abilene, TX 79602
(915) 677-4371

Wetsall Seed
PO Box 791
Harrisonburg, VA 22801
(703) 434-6753

Southern States Coop
Central Office
6606 W. Broad St.
Richmond, VA 23230
(804) 281-1353

TROPHY Brand Ladino Clover Mixture
TROPHY Brand *Plus* Ladino Clover/Alfalfa Mixture

Cal/West Seeds
PO Box 1428
Woodland, CA 95776
(800) 824-8585
Call for location of the nearest distributor.

MEGABUCKS clover seed mix

USAnalytic, Inc.
Box 606
Fogelsville, PA 18051
(800) 634-2822
Call for location of the nearest distributor.

No-Till seed mixtures

Wildlife Seed Blend
681 Tipton Rd.
Dade City, FL 33525
(800) 552-1027

IMPERIAL WHITETAIL Brand clover mixture

Whitetail Institute of North America
Rt 1 Box 3006
Pintlala, AL 36043
(800) 688-3030

Lablab

Tecomate Seed Co.
Rt. 2, Box 77A
San Juan, TX 78589
(210) 702-0870

Southeast Wildlife Mix

Lambert Seed & Supply, Inc.
PO Box 128
Camden, AL 36726
(205) 682-4111

Aquatic and Wetland Plants
Wildlife seed mixtures
Wildflower mixtures

Kester's Wild Game Food Nurseries, Inc.
PO Box 516
Omro, WI 54963
(800) 558-8815

Wildlife Mixes

Turner Seed Co.
Rt. 1 Box 292
Breckenridge, TX 76424
(817) 559-2065

Adams-Briscoe Seed
Company
P.O. Box 18
Jackson, GA 30233
(404) 775-7826

Anton Seed Co.
Lockhart, TX
(512) 398-2433

Wildflower Seed and Native Plants

Desert Enterprises
P.O. Box 23
Morristown, AZ
85342
(602) 388-2448

Wild Seed
P.O. Box 27751
Tempe, AZ 85285
(602) 345-0669

Carter Seeds
475 Mar Vista Dr.
Vista, CA 92083
1-800-872-7711

Santa Barbara Botanic
Garden
1212 Mission Canyon Rd.
Santa Barbara, CA 93105
(805) 682-9726

Theodore Payne
Foundation
10459 Tuxford St.
Sun Valley, CA 91352
(818) 768-1802

Valley Seed Service
P.O. Box 9335
Fresno, CA 93791
(209) 435-2163

Applewood Seed Co.
5380 Vivian St.
Arvada, CO 80002
(303) 431-6283

Arkansas Valley Seed Co.
4625 Colorado Blvd.
Denver, CO 80216
(303) 320-7500

Western Native Seed
P.O. Box 1281
Canon City, CO 81215
(719) 275-8414

Gann's Native Tropical
Greenery
22140 SW 152nd Ave.
Goulds, FL 33170
(305) 248-5529

Adams Briscoe Seed Co.
P.O. Box 18
Jackson, GA 30233
(404) 775-7826

Classic Groundcovers, Inc.
405 Belmont Rd.
Athens, GA 30605
(706) 543-0145

Northplan Seed Producers
P.O. Box 9107
Moscow, ID 83843
(208) 882-7446

Winterfeld Ranch Seed
Box 97, Hwy. 31 118
Swan Valley, ID 83449
(208) 483-3683

Seeds Trust
High Altitude Gardens
P.O. Box 1048
Hailey, ID 83333
(208) 788-4363

Country Road Greenhouses, Inc.
Rt. 1, Box 62
Malta, IL 60150
(815) 825-2305

Genesis Nursery
P.O. Box 342
Ladd, IL 61329
(815) 894-3329

Lafayette Home Nursery, Inc.
#1 Nursery Lane
Lafayette, IL 61449
(309) 995-3311

Midwest Wildflowers
Box 64
Rockton, IL 61072

Sand Prairie Farms
185 Rogers Rd.
Ohio, IL 61349
(815) 376-6641

Earl May Corporate Sales
208 North Elm
Shenandoah, IA 51603
1-800-843-9608

Hadfield Prairie Seed
Rt. 1, Box 32
McClelland, IA 51548
(712) 484-3326

Ion Exchange
Rt. 1, Box 48
Harpers Ferry, IA 52146
(319) 535-7231

Sharp Bros. Seed Co.
P.O. Box 140
Healy, KS 67850
(316) 398-2220

Shooting Star Nursery
444 Bates Rd.
Frankfort, KY 40601
(502) 223-1679

Environmental Concern, Inc.
P.O. Box P
St. Michaels, MD 21663
(410) 745-3517

Native Seeds, Inc.
14590 Triadelphia Mill Rd.
Dayton, MD 21036
(301) 596-9818

Hartmann's Plantation, Inc.
P.O. Box E
310 60th St.
Grand Junction, MI 49056
(616) 253-4281

International BioEnt Enterprises
14174 Hoffman Rd.
Three Rivers, MI 49093
(616) 278-3075

Landscape Alternatives, Inc.
1465 Pascal St.
St. Paul, MN 55108
(612) 488-3142

Prairie Hill Wildflowers
Rt. 1, Box 191-A
Ellendale, MN 56026
(507) 451-7791

Prairie Moon Nursery
Rt. 3, Box 163
Winona, MN 55987
(507) 452-1362

Prairie Restorations, Inc.
P.O. Box 327
Princeton, MN 55371
(612) 389-4342

Hamilton Seeds and Wildflowers
HCR 9, Box 138
Elk Creek, MO 65464
(417) 967-2190

Lawyer Nursery, Inc.
950 Hwy. 200 West
Plains, MT 59859
(406) 826-3881

Miller Seed Co.
1600 Cornhusker Hwy.
Lincoln, NE 68521
(402) 438-1232

Stock Seed Farms, Inc.
28008 Mill Rd.
Murdock, NE 68407
(402) 867-3771

Pinelands Nursery
323 Island Rd.
Columbus, NJ 08022
(609) 298-6846

Thompson & Morgan, Inc.
Dept. 102-4
P.O. Box 1308
Jackson, NJ 08527
(908) 363-9356

Wildland & Native Seeds Federation
2402 Hoffman Dr. NE
Albuquerque, NM 87110

Panfield Nurseries, Inc.
322 Southdown Rd.
Huntington, NY 11743
(516) 427-0112

Gardens of the Blue Ridge, Inc.
P.O. Box 10 Hwy. 221
Pineola, NC 28662
(704) 733-2417

Russell Graham Purveyor of Plants
4030 Eagle Crest Rd. NW
Salem, OR 97304
(503) 362-1135

Siskiyou Rare Plant Nursery
2825 Cummings Rd.
Medford, OR 97501
(503) 772-6846

Ernst Crownvetch Farms
RR#5, Box 806
Meadville, PA 16335
(814) 425-7276

Park Seed Co.
Box 31
Greenwood, SC 29647
(803) 223-8555

The Sexaner Company
P.O. Box 58
Brookings, SD 57006

Wildseed, Inc.
1101 Campo Rosa Rd.
Eagle Lake, TX 77434
1-800-848-0078

Intermountain Seed Co.
445 South First E.
Ephraim, UT 84627
(801) 283-4383

Maple Leaf Industries
480 South 50 East
Ephraim, UT 84627
(801) 283-4701

Vermont Wildflower Farm
Route 7
Charlotte, VT 05445
(802) 425-3500

Frosty Hollow
Box 53
Langley, WA 98260
(206) 221-2332

Seeds, Inc./Plants of the Wild
P.O. Box 866
Tekoa, WA 99033
(509) 284-2848

Art Lonergan
3048 Paradise Dr.
West Bend, WI 53095
(414) 334-3812

Country Wetlands Nursery
and Consulting
P.O. Box 126
Muskego, WI 53150
(414) 679-1268

CRM Ecosystems/Prairie
Ridge Nursery
9738 Overland Rd.
Mt. Horeb, WI 53572
(608) 437-5254

Little Valley Farm
RR 3, Box 544
Spring Green, WI 53588
(608) 935-3324

Prairie Nursery Inc.
P.O. Box 306
Westfield, WI 53964
(608) 296-2741

Prairie Seed Source
P.O. Box 83
North Lake, WI 53064
(414) 673-7166

Reeseville Ridge Nursery
309 S. Main St.
P.O. Box 171
Reeseville, WI 53579
(414) 927-3291

Retzer Nature Center
W284 SI530 Rd. DT
Waukesha, WI 53188
(414) 521-5407

West Wisconsin Nursery
Rt. 4, Box 131
Sparta, WI 54656
(608) 272-3171

Wind River Seed
Rt. 1, Box 97
Manderson, WY 82432
(307) 548-3361

Appendix II

Soil and Forage Testing Laboratories

Alabama

Soil Testing Laboratory
118 Funchess Hall
Auburn University, AL
36849-5411
(205) 844-3958

Arizona

Inter Ag Services, Inc.
2643 E. University Drive
Suite 113
Phoenix, AZ 85034
(602) 273-7248

Soil, Water and Plant
Analysis Lab
Dept. of Soil and Water
Science
Shantz Boulevard #429
University of Arizona
Tucson, AZ 85721
(602) 621-9703

Arkansas

Agriculture Diagnostic
Services Lab.
Plant Science 115
University of Arkansas
Fayetteville, AR 72701
(501) 575-3908

California

Atkins Farmlab
4017 Morehead Avenue
Chico, CA 95928
(916) 343-4947

Monarch Laboratory
563 E. Lindo Avenue
Chico, CA 95926
(916) 343-5818

Clarkson Laboratory &
Supply Inc.
350 Trousdale Drive
Chula Vista, CA 91910
(619) 425-1993

DANR Analytical
Laboratory
Hoagland Hall
University of California
Davis, CA 95616
(916) 752-0147

The National Food
Laboratory
6363 Clark Avenue
Dublin, CA 94568
(510) 828-1440

Wallace Laboratories
365 Coral Circle
El Segundo, CA 90245
(213) 615-0116

Dellavalle Laboratory, Inc.
1910 W. McKinley #110
Fresno, CA 93728-1298
(209) 233-6129

VPN Laboratory
3435 W. Shaw #104
Fresno, CA 93711
(209) 276-0401

J M Lord Inc.
267 N. Fulton
Fresno, CA 93711
(202) 268-9755

Western Agricultural Laboratory
416 E. 16th Street
P.O. Box 26
Hanford, CA 93232
(209) 582-9614

A&L Western Agricultural Laboratories
1311 Woodland Avenue #1
Modesto, CA 95351
(209) 529-4080

ALM Analytical Laboratory
77-725 Enfield Lane
Suite 110
Palm Desert, CA 92260
(619) 360-3050

Montgomery Laboratories
555 E. Walnut
Pasadena, CA 91101
(818) 568-6500

Harvey Laboratories
P.O. Box 686
205 D Street
Patterson, CA 95363-0686
(209) 892-9661

Environmental Technical Services
1343 Redwood Way
Petaluma, CA 94954
(707) 795-9605

Edward S. Babcock and Sons, Inc.
3215 Chicago Ave.
Riverside, CA 92507
(714) 684-1881

Quality Assurance Laboratory
6605 Nancy Ridge Dr.
San Diego, CA 92121
(619) 552-3636

Anresco, Inc.
1370 Van Dyke Avenue
San Francisco, CA 94124
(415) 822-1100

Coast-to-Coast Analytical Services
141 Suburban Road, C3
San Luis Abispo, CA 93401
(805) 549-0523

Soil and Plant Laboratory, Inc.
352 Mathew Street
Santa Clara, CA 95050
(408) 727-0330

Fruit Growers Laboratory
P.O. Box 272
853 Corporation Street
Santa Paula, CA 93061
(805) 659-0910

Truesdail Laboratories, Inc.
14201 Franklin Avenue
Tustin, CA 92680
(714) 730-6239

Branciforte Laboratories
46 Brennan Street
Watsonville, CA 95076
(408) 761-8478

Pajaro Valley Soil Laboratory
3 Parkwood Drive
Watsonville, CA 95076
(408) 722-5510

Colorado

Weld Laboratories, Inc.
1527 1st Avenue
Greeley, CO 80631
(303) 353-8118

Triple S Lab, Inc.
P.O. Box 678
Loveland, CO 80539
(303) 667-5671

Delaware

University of Delaware Soil Testing Lab
149 Townsend Hall
Newark, DE 19717-1303
(302) 451-1392

Florida

Pioneer Laboratory
4100 Glades Road
Ft. Pierce, FL 34981
(407) 464-9300

IFAS, Extension Soil Testing Laboratory
University of Florida
Wallace Bldg.
Gainesville, FL 32611
(904) 392-1950

Agro Services International, Inc.
215 E. Michigan Avenue
Orange City, FL 32763
(904) 775-6601

Joyce Environmental Consultants, Inc.
5051 North Lane
Orlando, FL 32808
(407) 297-7980

Thorton Laboratories, Inc.
1145 E. Cass Street
Tampa, FL 33602
(813) 223-9702

Georgia

Agricultural Services
Laboratories
2400 College Station Road
Athens, GA 30605
(404) 542-5350

Micro-Macro International,
Inc.
183 Paradise Boulevard
Suite108
Athens, GA 30607
(404) 548-4557

Waters Agricultural
Laboratories, Inc.
P.O. Box 382
Camille, GA 31730-0382
(912) 336-7216

Idaho

Western Laboratories, Inc.
P.O. Box 1020
Parma, ID 83660
(208) 722-6564

Illinois

Alvey Laboratory
1511 East Main Street
P.O. Box 175
Belleville, IL 62220
(618) 233-0445

Shields Soil Testing
Laboratory
RR. 1, Box 385
Dewey, IL 61840
(217) 897-1155

Key Agricultural Services,
Inc.
114 Shady Lane
Macomb, IL 61455
(309) 833-1313

Mowers Soil Testing Plus,
Inc.
117 East Main
P.O. Box 518
Toulon, IL 61483
(309) 286-2761

Indiana

Agri-Labs, Inc.
204 East Plymouth Street
Bremen, IN 46506
(219) 546-2606

A&L Great Lakes
Laboratories, Inc.
3505 Conestoga Drive
Ft. Wayne, IN 46808
(219) 483-4759

Iowa

Iowa Testing Laboratories,
Inc.
Hwy. #17, North
P.O. Box 188
Eagle Grove, IA 50533
(515) 448-4741

Kansas

Servi-Tech Laboratories
P.O. Box 1397
1816 East Wyatt Earp Blvd.
Dodge City, KS 67801
(316) 227-7123

Kentucky

UK Soil Testing Lab
103 Regulatory Services Bldg.
Lexington, KY 40546-0275
(606) 257-7355

McCoy & McCoy Laboratories, Inc.
85 East Noel Ave.
P.O. Box 907
Madisonville, KY 42431
(502) 821-7375

Michigan

Michigan State University Soil Testing Lab.
81 Plant & Soil Sciences Bldg.
Michigan State University
East Lansing, MI 48824
(517) 355-0218

Minnesota

Soil Testing & Research Analytical Labs
1903 Hendon Avenue
University of Minnesota
St. Paul, MN 55108
(612) 625-3101

Central (Agrisource) Laboratories
P.O. Box 64089, MS 370
St. Paul, MN 55164
(612) 451-4352

Mississippi

Extension Soil Testing Laboratory
P.O. Box 5446
Mississippi State, MS 39762
(601) 325-3313

Missouri

Missouri Extension Soil Testing Lab
21 Mumford Hall
University of Missouri
Columbia, MO 65211

Mey Soil Testing Lab
ET 122A, 4525 Downs Dr.
St. Joseph, MO 64507
(816) 271-4414

Montana

Energy Laboratories, Inc.
P.O. Box 30916
Billings, MT 59107
(406) 252-6325

Soil Analytical Laboratory
807 Leon Johnson Hall
Bozeman, MT 59717
(406) 994-5683

Western Testing Laboratory, Inc.
1920 9th Ave, North
P.O. Box 3165
Great Falls, MT 59403
(406) 761-1724

Nebraska

Servi-Tech Laboratories
P.O. Box 169
Hastings, NE 68902-0169
(402) 463-3522

Ward Laboratories, Inc.
P.O. Box 788
4007 Cherry Ave.
Kearney, NE 68848
(308) 234-2418

Harris Laboratory
624 Peach Street
Lincoln, NE 68501
(402) 476-2811

Olsen's Agricultural Laboratory, Inc.
P.O. Box 370
McCook, NE 69001-0370
(308) 345-3670

A&L Mid West Labs Inc.
13611 B Street
Omaha, NE 68144
(402) 334-7770

New Hampshire

Analytical Services Lab
Nesmith Hall
Durham, NH 03824
(603) 862-3212

New York

Cornell Nutrient Analysis Laboratory
803 Bradfield Hall
Soil, Crop and Atmospheric Sciences Dept.
Ithaca, NY 14853
(607) 255-1722

Fisons Analytical Laboratories
177 Sanfordville Road
Warwick, NY 10990
(914) 986-0447

North Carolina

Agronomic Division
NC Department of Agriculture
P.O. Box 27647
Raleigh, NC 27611
(919) 733-2655

Ohio

AG Analytical Services
2622 Baty Road
Elida, OH 45807
(419) 331-4550

Balance Labs
421 Leader Street
Marion, OH 43302
(614) 382-5708

Na-Churs Plant Food Company
421 Leader Street
Marion, OH 43302
(614) 382-5701

Brookside Laboratory, Inc.
308 South Main Street
New Knoxville, OH 45871
(419) 753-2448

Geophyta
2685 C.R. 254
Vickery, OH 43464
(419) 547-8538

Research-Extension
Analytical Laboratory
O.A.R.D.C.
Wooster, OH 44691
(216) 263-3760

Oregon

Columbia Laboratories, Inc.
36740 East Crown Point
Hwy.
P.O. Box 40
Corbett, OR 97019
(503) 695-2287

Central Analytical
Laboratory
Department of Crop & Soil
Science
Oregon State University
Corvallis, OR 97331
(503) 737-2178

Soils Testing Lab, Inc.
130 West Ninth
Medford, OR 97501
(503) 779-2833

Coffey Laboratories, Inc.
12423 NE Whitaker Wat
Portland, OR 97230
(503) 254-1794

Agri-Check, Inc.
323 Sixth Street
P.O. Box 705
Umatilla, OR 97882
(503) 922-4894

Pennsylvania

Grace-Sierra Testing
Laboratory
6656 Grant Way
Allentown, PA 18106-9316
(215) 395-7104

Agricultural Analytical
Services Laboratory
Pennsylvania State Univ.
University Park, PA 16803

Eastern Laboratory Service
Association
517 North Georgia Street
York, PA 17404
(717) 846-4953

Tennessee

University of Tennessee Soil
and Forage Testing
Laboratory
P.O. Box 1100019 (5201
Marchant)
Nashville, TN 37222-0019
(615) 832-5850

A&L Analytical
Laboratories, Inc.
411 North Third Street
Memphis, TN 38105
(901) 527-2780

Texas

Texas Plant & Soil
Laboratory, Inc.
Route 7, Box 213-y
Edinburg, TX 78539
(512) 383-0739

Western Agricultural
Laboratories, Inc.
P.O. Box 64666
Lubbock, TX 79464
(806) 794-4888

A&L Plains Ag Labs
302 34th Street
Lubbock, TX 79408
(806) 763-4278

Soil Testing Laboratory
Stephen F. Austin State
University
P.O. Box 9020, SFA Station
Nacogdoches, TX 75962
(409) 468-4500

Utah

Soil, Plant and Water
Analysis Laboratory
Utah State University, UMC
4830
Logan, UT 84322-4830
(801) 750-2217

Vermont

Agricultural Testing
Laboratory
Hills Bldg.
University of Vermont
Burlington, VT 05405
(802) 656-3030

Virginia

Virginia Tech Soil Testing &
Plant Analysis Laboratory
Virginia Tech
Blacksburg, VA 24061
(703) 231-6893

Ag. Consulting Laboratory
Route 1, Box 232
Mt. Crawford, VA 22841
(703) 234-8008

A&L Eastern Agricultural
Laboratories, Inc.
7621 Whiteline Road
Richmond, VA 23237
(804) 743-9401

Washington

Northwest Agricultural
Consultants
2545 West Falls
Kennwick, WA 99336
(509) 783-7459

Soiltest Farm Consultants
2925 Wapato Drive
Moses Lake, WA 98837
(509) 765-1622

USAg Analytical Services
1320 A East Spokane Street
Pasco, WA 99301
(509) 547-3838

Cascade Analytical, Inc.
3019 G. S. Center Road
Wenatchee, WA 98801
(509) 462-1888

Wisconsin

Dairyland Laboratories, Inc.
217 East Main
Arcadia, WI 54612
(608) 323-2123

UW Soil and Plant Analysis Laboratory
5711 Mineral Point Road
Madison, WI 53705
(608) 262-4364

UW Soil and Forage Analysis Laboratory
8396 Yellowstone Drive
Marshfield, WI 54449

K. Ag Laboratories International, Inc.
2323 Jackson Street
Oshkosh, WI 54901
(414) 426-2222

Wyoming

Soil Testing Laboratory
Dept. Plant & Soil Sciences
P.O. 3354, University Station
Laramie, WY 82071
(307) 766-2135

Inter-Mountain Laboratories, Inc.
1633 Terra Ave.
Sheridan, WY 82801

Notes

Notes

Notes

Notes